Computational Overview of Fluid Structure Interaction

Edited by Khaled Ghaedi, Ahmed Alhusseny, Adel Nasser and Nabeel Al-Zurfi

Published in London, United Kingdom

IntechOpen

Supporting open minds since 2005

Computational Overview of Fluid Structure Interaction
http://dx.doi.org/10.5772/intechopen.92949
Edited by Khaled Ghaedi, Ahmed Alhusseny, Adel Nasser and Nabeel Al-Zurfi

Contributors
Sukhmander Singh, Michael Roberts, Mayra Karina Zezatti Flores, Laura Castro Gómez, Gustavo Urquiza, Nahom Alemseged Alemseged Worku, Adnène Arbi, Assad Ayub, Hafiz A. Wahab, Zulqurnain Sabir, Nguyen Minh Phu, Nguyen Van Hap, Deepak Srivastava, Khaled Ghaedi, Farzad Hejazi, Meisam Gordan, Ahad Javanmardi, Hamed Khatibi, Ali Joharchi

Notice
Statements and opinions expressed in the chapters are these of the individual contributors and not necessarily those of the editors or publisher. No responsibility is accepted for the accuracy of information contained in the published chapters. The publisher assumes no responsibility for any damage or injury to persons or property arising out of the use of any materials, instructions, methods or ideas contained in the book.

First published in London, United Kingdom, 2021 by IntechOpen
IntechOpen is the global imprint of INTECHOPEN LIMITED, registered in England and Wales, registration number: 11086078, 5 Princes Gate Court, London, SW7 2QJ, United Kingdom
Printed in Croatia

British Library Cataloguing-in-Publication Data
A catalogue record for this book is available from the British Library

Additional hard and PDF copies can be obtained from orders@intechopen.com

Computational Overview of Fluid Structure Interaction
Edited by Khaled Ghaedi, Ahmed Alhusseny, Adel Nasser and Nabeel Al-Zurfi
p. cm.
Print ISBN 978-1-83969-175-1
Online ISBN 978-1-83969-176-8
eBook (PDF) ISBN 978-1-83969-177-5

For EU product safety concerns:
IN TECH d.o.o., Prolaz Marije Krucifikse Kozulić 3, 51000 Rijeka, Croatia, info@intechopen.com or visit our website at intechopen.com.

Meet the editors

Dr. Khaled Ghaedi has been in both the academia and industry fields. He has built his professional career holding different positions over the years with experience in engineering and research. He is the technical director of the engineering software and solutions provider in Malaysia to support the industry in Southeast Asia. He is the committee member of the international structural design code, ASCE 7-22. He has frequently published his research papers in well-known journals in different engineering fields such as earthquake engineering, finite element analysis, computational fluid dynamics, and materials and damage assessment of buildings, bridges, and dams. He has contributed to the process of knowledge sharing through his roles as an editorial board member, reviewer, and committee member of eminent journals, conferences institutes, and organizations.

Ahmed Alhusseny received a BSc and MSc in HVAC&R Engineering from the University of Technology, Iraq, in 2001 and 2005, respectively. He then joined the University of Kufa, Najaf, Iraq, as a member of academic staff in the Department of Mechanical Engineering. Later, he joined the School of MACE, University of Manchester, UK, as a researcher where he was awarded a Ph.D. in Mechanical Engineering in 2016. Dr. Alhusseny is still associated with the University of Manchester as a researcher although he has returned to Iraq and resumed his former job there as an assistant professor in the Department of Mechanical Engineering and the head of the Branch of Air-conditioning and Refrigeration Engineering at the University of Kufa.

Dr. Adel Nasser is currently a senior lecturer at the University of Manchester. He developed and delivered several undergraduate and postgraduate modules on computing, modeling, and simulations. His research revolves around the use of state-of-the-art modeling and simulation techniques to tackle a wide range of topics within fluid dynamics and solid mechanics and their interaction in complex engineering problems. This covers renewable energy (both solar and wind energy), turbulence, convective heat transfer, multiphase flows, and thermal energy storage technologies. Such research contributes to the blueprint of more efficient, economically affordable energy sources to improve the lifestyle of society and individuals alike. Dr. Nasser has extensive experience in IT services involving frontline service, high-performance computing, and eLearning systems. www.research.manchester.ac.uk/portal/a.g.nasser.html

Nabeel Al-Zurfi obtained a BSc and MSc in Mechanical Engineering from the University of Kufa, Najaf, Iraq, in 2002 and 2005, respectively. He then worked as a member of academic staff in the Department of Mechanical Engineering at the same university. Later, he joined the School of MACE, University of Manchester, UK, as a sponsored Ph.D. student where he was awarded a Ph.D. in Mechanical Engineering in 2016. Dr. Al-Zurfi is still associated with the University of Manchester as a visiting researcher although he has returned to Iraq and resumed his former job there as an assistant professor in the Department of Mechanical Engineering, University of Kufa.

Contents

Preface

Fluid-Structure Interaction (FSI), also known as engineering fluid mechanics, deals with mutual interaction between fluid and structural components. The fluid flow, depending on the structural shape, motion, surface, and structural roughness, acts as mechanical forces on the structure. FSI can be seen everywhere in medicine, engineering, aerospace, the sciences, and even our daily life, for example, when a water tap is opened or blocked. This book provides the basic concept of fluid flow behavior in interaction with structures, which is crucial for almost all engineering disciplines.

Along with the fundamental principles, the book covers a variety of FSI problems ranging from fundamentals of fluid mechanics to plasma physics, wind turbines and their turbulence, heat transfer, magnetohydrodynamics, and dam-reservoir systems. This extensive and methodically prepared book provides a concise, detailed, and thorough discussion of the essentials and principles in FSI. The chapters cover the theory and study of fluid mechanics with a focus on problems concerning fluid flow using simple mathematical expressions.

Each chapter starts with statements of relevant explanations, definitions, and axioms together with instructive material for solving the problems. The solved problems show the theory and recommend computation approaches to empower the reader to implement the fundamental rules properly and with confidence. Derivations of theoretical formulations are involved within the explained problems and the discussed materials in each chapter serve as a comprehensive review of the FSI problems.

This book is a useful reference for fluid mechanics students as well as practicing engineers and we hope our readers benefit from the information contained herein. We would be pleased to receive comments, suggestions, and/or criticisms.

Dr. Khaled Ghaedi
Head of Center of Research and Development,
PASOFAL Engineering Group,
Kuala Lumpur, Malaysia

Head of Research and development division,
Hormoz Beton Co.,
Bandar Abbas, Iran

Senior Researcher,
Department of Civil Engineering, University of Malaya,
Kuala Lumpur, Malaysia

Technical Director,
Division of Engineering Software solutions,
VR-CAM Technologies Sdn Bhd,
Kuala Lumpur, Malaysia

Ahmed Alhusseny, Adel Nasser and Nabeel Al-Zurfi
University of Manchester,
United Kingdom

Chapter 1

Basics of Fluid Dynamics

Nahom Alemseged Worku

Abstract

In this chapter, studies on basic properties of fluids are conducted. Mathematical and scientific backgrounds that helps sprint well into studies on fluid mechanics is provided. The Reynolds Transport theorem and its derivation is presented. The well-known Conservation laws, Conservation of Mass, Conservation of Momentum and Conservation of Energy, which are the foundation of almost all Engineering mechanics simulation are derived from Reynolds transport theorem and through intuition. The Navier–Stokes equation for incompressible flows are fully derived consequently. To help with the solution of the Navier–Stokes equation, the velocity and pressure terms Navier–Stokes equation are reduced into a vorticity stream function. Classification of basic types of Partial differential equations and their corresponding properties is discussed. Finally, classification of different types of flows and their corresponding characteristics in relation to their corresponding type of PDEs are discussed.

Keywords: Navier–Stokes equation, engineering simulation, conservation laws, Reynolds transport equation, modeling, simulation, fluid dynamics, fluid flow, nonlinear differential equations

1. Introduction

1.1 Mathematical background

In this subsection, important mathematical formulations and ideas that help understand fluids are discussed.

All mathematical formulations presented in this chapter will be performed using vectors. Therefore, it is advised to have a good knowledge of the fundamental theorems of vector calculus.

1.1.1 Green's theorem

Assuming a closed curve **C**, the Green's theorem expresses contour integral of **C** in terms of a two dimensional region **R,** bounded by **C** [1–3].

The Green's theorem is given as

$$\int_R \left[\frac{\partial Q}{\partial x} - \frac{\partial P}{\partial y}\right] dA = \oint_c P dx + Q dy \tag{1}$$

1.1.2 Stokes theorem

Let Q be a vector field, s be an oriented surface be a closed surface oriented by the Right hand rule, Stokes theorem states that

$$\int_{s} [\nabla \times \mathbf{Q}] \boldsymbol{ds} = \oint \boldsymbol{Qdr} \tag{2}$$

where r is such that dr/ds is the unit tangent vector and s the arc length of C. The curve of the line integral, C, must have positive orientation, meaning that dr points counterclockwise when the surface normal, dS, points toward the viewer, as per the right-hand rule [1–3].

1.1.3 Divergence theorem

Divergence theorem is a relation to convert volume integral into areal integral.

Let v be a volume in a three dimensional space, and $\boldsymbol{\Omega}$ be the surface boundary. Let $\boldsymbol{\eta}$ be a unit normal pointing outward from the surface. **Q** being any vector field, the Divergence theorem is given as

$$\int_{v} Q.vdv = \int_{\Omega} Q.\eta dA \tag{3}$$

The Divergence theorem is a very important concept, as shall be discussed in later sections, in the area of fluid dynamics, especially in studying the Flux terms [1–6].

The Divergence Theorem is equally applicable to tensors.

1.1.4 Leibniz integral rule

Leibniz integral rule gives a formula for differentiating a definite integral whose limits are a function of definite variable [1–6]. If Q be a field that is a function of time t and space X, the Leibniz Rule is given as

$$\frac{d}{dx}\int_{a(t)}^{b(t)} f(x,t)dx = \int_{a(t)}^{b(t)} \frac{\partial f}{\partial x}dx + b'(t)f(b(t),\mathrm{t}) - \mathrm{a}'(\mathrm{t})\mathrm{f}(\mathrm{a}(\mathrm{t}),\mathrm{t}) \tag{4}$$

If a and b are constant, the second and third terms go to zero.

1.2 Background on fluids

1.2.1 What are fluids?

Air is all around us. We drink water every day. We clean ourselves and our environment using water. Almost 71% of the earth is covered with water. Our lives are highly interrelated with fluids. This highly necessitates the study of fluids.

Fluids can be found in liquid or Gaseous state.

Fluids, in its Engineering sense, can be defined as a material that shear constantly in the presence of a very small disturbance (Force and/or Gradient). Assume we pour water over a horizontal plate. The water will flow horizontally in all directions even if there is no gradient applied until it reaches stable position of a very minimal depth.

The study of fluid dynamics is a very important area and is very useful in the area of modeling and simulation of fluid flow.

1.2.2 Scale

Scale is a very important concept when studying a natural phenomenon.

It helps understand where to position oneself to look at his/her study. There are two major categories of scales, namely, microscopic and macroscopic scales.

If we see water with our naked eye, it is continuous and smooth, i.e. macro scale. But when looked under microscope, we see small discontinuity. Zooming it a bit more, it becomes more discontinuous. It somehow looks like a dense crowd in a subway. Again zooming it more, we see groups (large chunks) of circles grouped together and moving along with each other. Finally, if we zoom it enough, we can see groups with three circles joined and moving together. The three circles joined together are water molecules (H2O), with two hydrogen atoms and one oxygen atom (micro scale).

The scale below molecular level, i.e., molecules, atoms, subatomic particles are microscopic scale. The scale above which can be seen with the naked eye is commonly called macroscopic scale. In fluid mechanics, and also in solid mechanics, macroscopic level of study is performed. In fluid mechanics, as in the case of solid mechanics, materials are continuous and are thought of being composed of macroscopic elements (chunks). A chunk of fluid and solid, called a control volume, is used to study the overall property of fluids and solids respectively.

1.2.3 Frame of reference

In engineering mechanics, there are three types of frame of reference. The lagrangian, Eulerian and the Arbitrary-Lagrangian–Eulerian frame of reference [7].

- **Lagrangian Frame of Reference (L.F.R):**- In the lagrangian frame of reference, properties of material points are studied by tracing individual material elements. Let us assume there is a hypothetical grid of reference aligned with the material element. In the L.F.R., the reference grid is not stationary and deforms together with the domain.

  In fluids, the L.F.R. study can be implemented by using streak lines (dyes). The movement of the dye in the fluid is assumed to be a particle of fluid to be studied. There are some cases that the lagrangian approach can be used. In cases of solid mechanics, since the particles undergo very small deformations, we can allow the reference grid to deform along with the body. Hence, the Lagrangian frame of reference is preferred.

- **Eulerian Frame of Reference(EFR):**- In the EFR case, we assume a stationary reference grid to monitor properties at a specific point and time. Normally, in fluids, it is very difficult and can be unnecessary to track infinitely many fluid particles. Hence, the Lagrangian frame of reference cannot be used.

  Instead, the Eulerian frame of reference studies properties of fluids at a specific space and time, which makes it convenient to study fluids. It is performed by tracing properties of the fluid at each stationary grid point.

- **Arbitrary Eulerian Lagrangian Frame of Reference (ALE):**

  We have seen that in case of the Lagrangian Frame of Reference, the reference grid moves independently and in the case of the Eulerian Frame of Reference, the grid is stationary.

In the case of the Arbitrary Eulerian Lagrangian Frame of Reference, the reference grid moves independently with the material element. This type of frame of reference is called the ALE. The ALE frame of reference is widely used in the study of fluid–structure Interaction problems.

1.2.4 Types of flow

1.2.4.1 Laminar and Turbulent Flow

Flows can be broadly classified as Laminar and Turbulent flows. At a certain velocity and viscosity, flows are stable and have a defined property. The viscosity tend to dissipate the velocity and pressure terms and take the responsibility of calming the flow that it has a defined property, which is termed as **Laminar flow** [1, 8].

But when the viscosity is small as compared to that of velocity of the flow, the flow shall have unpredictable and chaotic property, thereby called **Turbulent flow** [1, 8].

The relation between the velocity and viscosity of a flow can be described by a dimensionless number called Reynold's Number, which is given as the ratio of Velocity times a length scale to that of viscosity.

$$\text{Re} = \frac{\rho U l}{\mu} \tag{5}$$

Where Re is the Reynold's Number, ρ is the density of the fluid, U is the velocity, l is the Length scale, and μ is the dynamic Viscosity.

As the Re is below a certain value for a specific value for a certain flow, the flow is classified as a Laminar Flow, and is classified as Turbulent when Re is higher than the stipulated value for the specific type of flow.

1.2.4.2 Viscous and Inviscid Flow

A fluid is formally termed viscous if the shear stress is directly proportional to the shear strain rate. In solids, materials with stress directly proportional to strain are called Elastic material, or said to obey Hook's Law.

A fluid is inviscid if shear stress is not directly proportional to shear strain rate. By the same token, solids are termed as Inelastic if stress and strain are not directly related.

Viscous fluids exhibit nonlinear behavior. This can intuitively be demonstrated by the feel someone will have if s/he spills a ketchup or oil and water. Water is relatively in viscid while oil or ketchup is viscous.

Viscosity dissipates energy thereby, stabilizes a fast moving Laminar flow [1, 2, 8–12]. But, it can some destabilize a flow more in some Turbulent flow.

While viscosity dissipates energy, elastic materials store energy.

2. Conservation laws

In fluid mechanics, for convenience, individual fluid particles are called fluid particles. And a set of fluid particles comprise a fluid element or a fluid system. Therefore, one fluid element can comprise of many fluid particles, as can be seen in **Figure 1**.

To find answers to different physical problems that arise in fluids, we can simply apply the fundamental laws of physics. But, the problem is that, physical laws are obtained in the Lagrangian frame of reference form. Therefore, it is necessary to customize it to the Eulerian frame of reference. To do that, we use the famous the Reynolds Transport Theorem.

2.1 Reynold's transport equation

Many of Fluid Dynamics problems are of interest in understanding and solving what is currently going on at a specific point and time, rather than tracking particles (Eulerian rather than Lagrangian Study).

But unfortunately, Physical laws like Newton's laws can be applied in the Lagrangian Frame of Reference (**Figure 2**).

The Reynolds transport theorem converts Eulerian Study into the Lagrangian one so that the physical laws can be customized to the Eulerian frame of reference.

The above figure shows a material element (fluid element) in a motion. At time t, the fluid element was at position 1. And after a time increase of Δt,i.e. at time $t + \Delta t$, it moves to position 3. In the moving process, the element t and $t + \Delta t$ intersected at position 2.

Now, let N be any arbitrary extensive property. Then,

$$N_t = (N_1)_t + (N_2)_t, \text{ and } N_{t+\Delta t} = (N_2)_{t+\Delta t} + (N_3)_{t+\Delta t}. \qquad (6)$$

Figure 1.
Schematic diagram for fluid particle and fluid element.

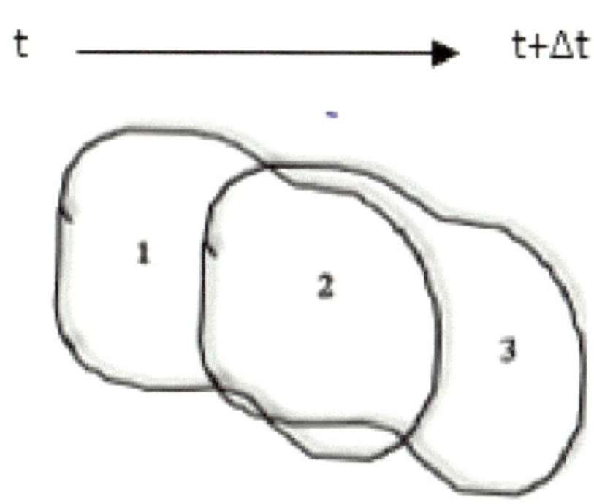

Figure 2.
Schematic of fluid particle motion to visualize Reynolds transport.

The rate of change of property N with respect to time t is given as

$$\frac{dN}{dt} = \lim_{\Delta t \to 0} \frac{N_{t+\Delta t} - N_t}{\Delta t}, \tag{7}$$

Solving Eq. (2), we can obtain

$$\frac{dN}{dt} = \lim_{\Delta t \to 0} \frac{(N_2)_{t+\Delta t} - (N_2)_t}{\Delta t} + (N_3)_{t+\Delta t} - (N_1)_t \tag{8}$$

Here, region 2 is our control volume. And region 1 is property ready to enter the control volume and region 3 is a region leaving the control volume [7].

In Eq. (4), the first term on the right hand side is the rate of *change of material property N with respect to time. The second term is the outflow from the control volume and the third term is the inflow to the control volume.

Therefore, Eq. (4) can be read as the rate if change of a material property N with respect to time t is equal to the rate of increase of N in the control volume plus the net flux i.e. Outflow minus the inflow rate of the system.

Rate of change of N = Net rate of change w.r.t. time + Net flux into and out of C.V.

Net flux >0 if inflow is less than outflow and.

Net flux <0 if inflow is greater than outflow.

Now, let us introduce a derived property Φ, which is the rate of N per unit mass.

$$\mathrm{N} = \Phi * \mathrm{m}.$$

Flux is written in terms of control surface instead of control volume. Therefore, using divergence theorem, the net flux is given by

$$\int_{cs} \rho\phi u.nds \tag{9}$$

Where u is the velocity of the fluids and n is the outward unit normal.

The outward unit normal is normal to the surface since only the normal components of the flux terms enter and leave the control volume.

Hence, the general transport equation of extensive property Φ is

$$\frac{d}{dt}\int_{v} \rho\phi dv = \frac{\partial}{\partial t}\int_{cv} \rho\phi dv + \int_{cs} \rho\phi u.nds \tag{10}$$

2.2 Conservation of mass

The conservation of mass states that mass of fluids in a system is conserved [1, 2, 8–12].

$$\text{Rate of Mass increase in a fluid element} = \text{Net rate of flow of mass in the fluid.}$$

So using the transport equation, we can derive the conservation of mass general equation.

Now, let the property N be mass m. Therefore, our desired property Φ be m/m which is equal to 1. Therefore, plugging this into Eq. (10) yields,

$$\frac{d}{dt}\int_{v} \rho dv = \frac{\partial}{\partial t}\int_{cv} \rho dv + \int_{cs} \rho u.nds \tag{11}$$

But, since mass is conserved, the term on the left hand side is zero, i.e. the net rate of change of mass is zero. Therefore,

$$\frac{\partial}{\partial t}\int_{cv}\rho dv + \int_{cs}\rho u.nds = 0 \tag{12}$$

To write Eq. (12) in compact form, the flux term can be written in the form of control volume, instead of control surface. Hence,

$$\int_{cs}\rho u.nds = -\int_{cv}\nabla.\rho udv \tag{13}$$

$$\frac{\partial}{\partial t}\int_{cv}\rho dv + \int_{cv}\nabla.\rho udv = 0 \tag{14}$$

Finally, the compact form of conservation of mass is given by

$$\int_{cv}\left[\frac{\partial\rho}{\partial t} + \nabla.\rho u\right]dv = 0 \tag{15}$$

Physical Intuition method of deriving the conservation of Mass Equation.

We describe the behavior of the fluid in terms of macroscopic properties, such as velocity, pressure, density and temperature, and their space and time derivatives. These may be thought of as averages over suitably large numbers of molecules. A fluid particle or point in a fluid is then the smallest possible element of fluid whose macroscopic properties are not influenced by individual molecules. We consider such a small element of fluid with sides δ x, δ y and δ z.

From **Figure 3**, we can see that there are six faces labeled as N, S, W, E, T and B. The positive directions are given in the figure (**Figure 4**).

We should notice that all properties are functions of space coordinates X, Y, Z and time component t.

The element under consideration is so small that fluid properties at the faces can be expressed accurately enough by means of the first two terms of a Taylor series expansion. Let N be an arbitrary material property, then N at the W and E faces, which are both at a distance of 1/2 δ x from the element center, can be expressed as

$$N - \frac{\partial N}{\partial x} * \frac{1}{2}\delta x \text{ and } N + \frac{\partial N}{\partial x} * \frac{1}{2}\delta x. \tag{16}$$

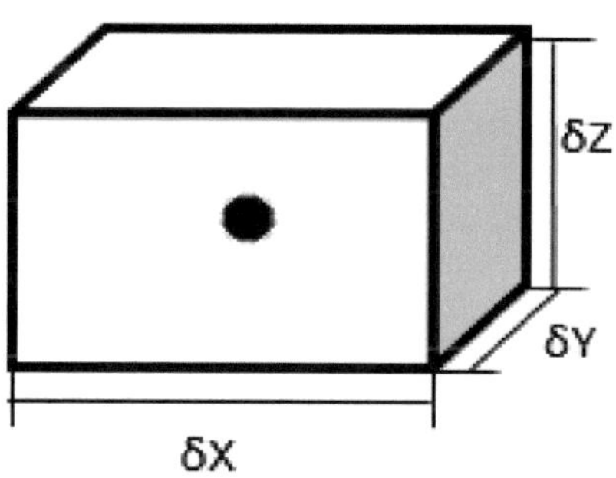

Figure 3.
Infinitesimal fluid element.

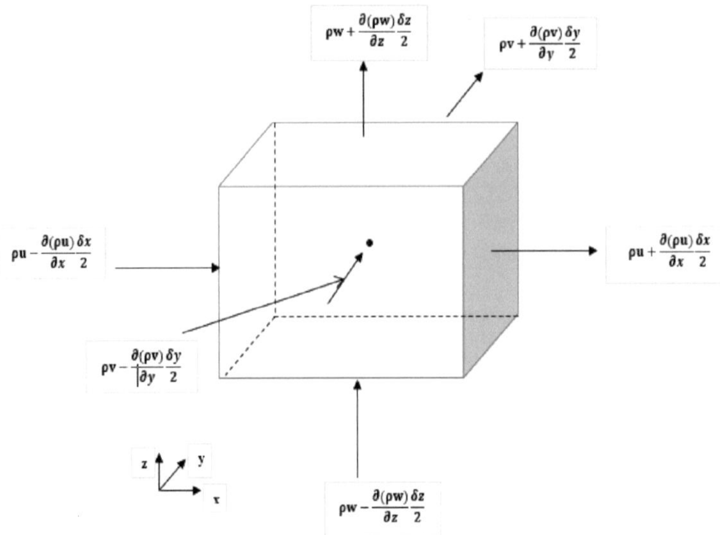

Figure 4.
Infinitesimal fluid element of mass transfer.

The net rate of increase of fluid element is given by

$$\frac{\partial m}{\partial t} = \frac{\partial(\rho dv)}{\partial t} \tag{17}$$

$$\frac{\partial(\rho dv)}{\partial t} = \frac{\partial \rho}{\partial t}(\delta x \delta y \delta z) \tag{18}$$

Assuming inflow to the fluid element to be positive and outflow to be negative. The net flow rate into and out of the fluid element is given as

$$\begin{aligned}&\left[\rho u - \frac{\partial(\rho u)}{\partial x}\frac{\delta x}{2}\right]\delta y \delta z - \left[\rho u + \frac{\partial(\rho u)}{\partial x}\frac{\delta x}{2}\right]\delta y \delta z \\ &+ \left[\rho v - \frac{\partial(\rho v)}{\partial y}\frac{\delta y}{2}\right]\delta x \delta z - \left[\rho v + \frac{\partial(\rho v)}{\partial y}\frac{\delta y}{2}\right]\delta x \delta z \\ &+ \left[\rho w - \frac{\partial(\rho w)}{\partial z}\frac{\delta z}{2}\right]\delta x \delta y - \left[\rho w + \frac{\partial(\rho w)}{\partial z}\frac{\delta z}{2}\right]\delta x \delta y\end{aligned} \tag{19}$$

Therefore, equating Eq. (18) and (19), we obtain

$$-\left[\frac{\partial(\rho u)}{\partial x} + \frac{\partial(\rho v)}{\partial y} + \frac{\partial(\rho w)}{\partial z}\right] = \frac{\partial \rho}{\partial t} \tag{20}$$

$$\frac{\partial \rho}{\partial t} + \left[\frac{\partial(\rho u)}{\partial x} + \frac{\partial(\rho v)}{\partial y} + \frac{\partial(\rho w)}{\partial z}\right] = 0 \tag{21}$$

$$\frac{\partial \rho}{\partial t} + \nabla.(\rho u) = 0 \tag{22}$$

Eq. (21) is an unsteady, three dimensional mass conservation or continuity equation.

In case of incompressible fluids like water, the density do not change with time and space and hence Eq. (35) can be reduced to

$$\nabla.\mathrm{u} = 0 \tag{23}$$

In the long hand notation, the equation becomes

$$\frac{\partial u}{\partial x} + \frac{\partial v}{\partial y} + \frac{\partial w}{\partial z} = 0 \tag{24}$$

Rates of change following a fluid particle and for a fluid element.

Let the value of a property per unit mass be denoted by Φ. The total or substantive derivative of Φ with respect to time following a fluid particle, written as D Φ/Dt, is

$$\frac{D\phi}{Dt} = \frac{d\phi}{dt} + \frac{d\phi}{dx}\frac{dx}{dt} + \frac{d\phi}{dy}\frac{dy}{dt} + \frac{d\phi}{dz}\frac{dz}{dt} \tag{25}$$

Here, dx/dt = u, dy/dt = v, dz./dt = w. and hence,

$$\frac{D\phi}{Dt} = \frac{d\phi}{dt} + u\frac{d\phi}{dx} + v\frac{d\phi}{dy} + w\frac{d\phi}{dz} \tag{26}$$

$$\frac{D\phi}{Dt} = \frac{d\phi}{dt} + u.grad(\phi) \tag{27}$$

Where u, v and w are velocity in the x, y, and z direction.

$\frac{D\phi}{Dt}$ Is a property that is defined as the property per unit mass. But, we are interested in developing equations of rates of change per unit volume. Therefore. By multiplying the term $\frac{D\phi}{Dt}$ by density, we can obtain rates of change per unit volumc.

Therefore,

$$\rho\frac{D\phi}{Dt} = \rho\left[\frac{d\phi}{dt} + u.grad(\phi)\right] \tag{28}$$

The generalization of these terms for an arbitrary conserved property is

$$\frac{\partial(\rho\phi)}{\partial t} + \nabla.\rho\phi u = 0 \tag{29}$$

The above equation Eq. (41) expresses the rate of change in time of Φ per unit volume plus the net flow of Φ out of the fluid element per unit volume. It is now rewritten to illustrate its relationship with the substantive derivative of Φ.

$$\begin{gathered}\frac{\partial(\rho\phi)}{\partial t} + \nabla.\rho\phi u = \rho\left[\frac{\partial(\phi)}{\partial t} + \nabla.\phi u\right] + \phi\left[\frac{\partial(\rho)}{\partial t} + \nabla.\rho u\right] \\ \nabla.\phi u = \frac{\partial(\phi u)}{\partial x} + \frac{\partial(\phi v)}{\partial y} + \frac{\partial(\phi w)}{\partial z} \\ \frac{\partial(\phi u)}{\partial x} + \frac{\partial(\phi v)}{\partial y} + \frac{\partial(\phi w)}{\partial z} = [u + v + w].\left[\frac{\partial(\phi)}{\partial x} + \frac{\partial(\phi)}{\partial y} + \frac{\partial(\phi)}{\partial z}\right] \\ \nabla.u\phi = u.\nabla\phi\end{gathered} \tag{30}$$

But since the second term on the right hand side is the conservation of mass equation which is zero, therefore

$$\rho\frac{D\phi}{Dt}=\frac{\partial(\rho\phi)}{\partial t}+\nabla.\rho\phi u=\rho\left[\frac{\partial(\phi)}{\partial t}+u.\nabla\phi\right] \tag{31}$$

2.3 Conservation of momentum

The conservation of Momentum states that the sum of the rate of change of momentum on a fluid particle is equal to the sum of forces on the particle. This is basically Newton's second law.

$$\text{Rate of change of Momentum On a fluid particle} \\ = \text{Sum of forces on a fluid particle}$$

Our property N is now momentum P. Therefore, P = m*u. Therefore, Φ is the velocity u since momentum is equal to mass times velocity.

From Reynolds's transport theorem, we can obtain

$$\begin{gathered}\frac{\partial}{\partial t}\int\limits_{cv}\rho u dv+\int\limits_{cs}\rho u.u.nds=0\\ \int\limits_{cv}\left[\frac{\partial(\rho u)}{\partial t}+\nabla.\rho uu\right]=0\end{gathered} \tag{32}$$

Eq. (32), which is the integral form is used for the fluid element.

For fluid particle, we can use the differential form as.

Table 1. Conservation equations in vector form

$$\left[\frac{\partial(\rho u)}{\partial t}+\nabla.\rho uu\right]=0 \tag{33}$$

But Eq. (33) deals with the rate of change of Momentum. Now we shall see the force components of the equation.

2.3.1 Types of forces on fluid particles

There are generally two types of forces on fluids.

Surface Forces: Are type of forces that are applied on surfaces (area). Some of the Surface forces pressure forces, viscous and the like [8].

Body Forces: Are type of forces that are applied on volumes. Some of the Body forces gravity forces, electromagnetic forces, centrifugal forces [1, 8].

There are nine viscous stress terms as state of stress and one pressure term as can be seen on **Figure 5**.

X-Momentum	U	$\rho\frac{Du}{Dt}$	$\left[\frac{\partial(\rho u)}{\partial t}+\nabla.\rho uu\right]$
Y-Momentum	V	$\rho\frac{Dv}{Dt}$	$\left[\frac{\partial(\rho v)}{\partial t}+\nabla.\rho vu\right]$
Z-Momentum	W	$\rho\frac{Dw}{Dt}$	$\left[\frac{\partial(\rho w)}{\partial t}+\nabla.\rho wu\right]$
Energy	E	$\rho\frac{DE}{Dt}$	$\left[\frac{\partial(\rho E)}{\partial t}+\nabla.\rho Eu\right]$

Table 1.
Vector notation of conservation Laws.

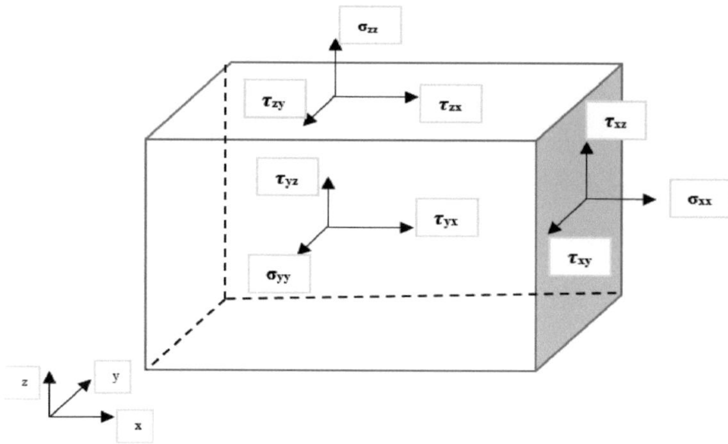

Figure 5.
Infinitesimal fluid element of momentum transfer with shear stress.

The pressure, a normal stress, is denoted by p. Viscous stresses are denoted by τ. The usual suffix notation τ_{ij} is applied to indicate the direction of the viscous stresses. The suffices i and j in τ_{ij} indicate that the stress component acts in the j direction on a surface normal to the I $_{th}$-direction [1, 8] (**Figure 6**).

Therefore, the summation of forces in the X direction is given below.

Summation for the Pressure term is given as

$$\left[\left[P-\frac{\partial P}{\partial x}\frac{\delta x}{2}\right]-\left[P+\frac{\partial P}{\partial x}\frac{\delta x}{2}\right]\right]\delta y\delta z \tag{34}$$

Eq. (46) can be reduced to

$$-\frac{\partial P}{\partial x}\delta x\delta y\delta z \tag{35}$$

Similarly, for the viscous shear stress along the X-direction at the east and west sides is given as

$$\left[\left[\tau_{xx}+\frac{\partial \tau_{xx}}{\partial x}\frac{\delta x}{2}\right]-\left[\tau_{xx}-\frac{\partial \tau_{xx}}{\partial x}\frac{\delta x}{2}\right]\right]\delta y\delta z \tag{36}$$

$$\frac{\partial \tau_{xx}}{\partial x}\delta x\delta y\delta z \tag{37}$$

Similarly, for the viscous shear stress along the X-direction at the North and South sides is given as

$$\left[\left[\tau_{yx}+\frac{\partial \tau_{yx}}{\partial y}\frac{\delta y}{2}\right]-\left[\tau_{yx}-\frac{\partial \tau_{yx}}{\partial y}\frac{\delta y}{2}\right]\right]\delta x\delta z \tag{38}$$

$$\frac{\partial \tau_{yx}}{\partial y}\delta x\delta y\delta z \tag{39}$$

Again for the viscous shear stress along the X-direction at the Top and Bottom sides is given as

$$\left[\left[\tau_{zx}+\frac{\partial \tau_{zx}}{\partial z}\frac{\delta z}{2}\right]-\left[\tau_{yx}-\frac{\partial \tau_{zx}}{\partial z}\frac{\delta z}{2}\right]\right]\delta x\delta y \tag{40}$$

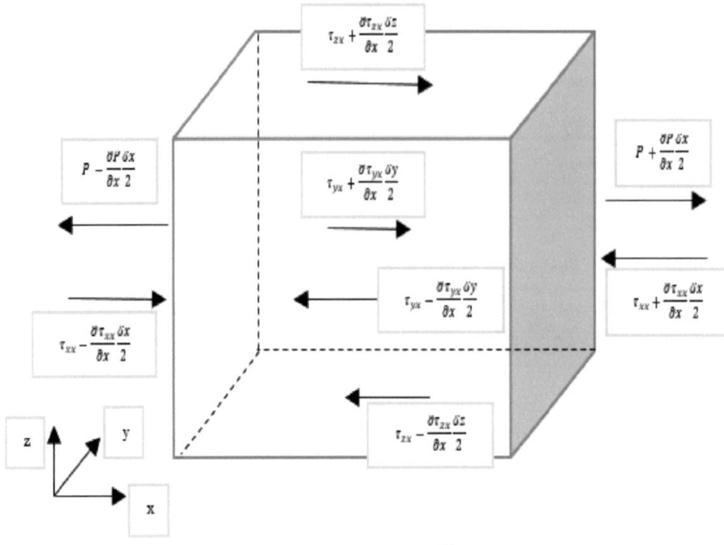

Figure 6.
Infinitesimal fluid element of momentum transfer with shear stress and pressure term.

$$\frac{\partial \tau_{zx}}{\partial z}\delta x \delta y \delta z \tag{41}$$

Therefore, the Total net force per unit volume along the X-axis is given as

$$\frac{\partial(-P + \tau_{xx})}{\partial x} + \frac{\partial \tau_{yx}}{\partial y} + \frac{\partial \tau_{zx}}{\partial z} \tag{42}$$

Finally, the General Conservation of Momentum Equation is given as

$$\frac{\partial(-P + \tau_{xx})}{\partial x} + \frac{\partial \tau_{yx}}{\partial y} + \frac{\partial \tau_{zx}}{\partial z} = \rho \frac{Du}{Dx} \tag{43}$$

And the Final equation for X-Momentum is

$$\frac{\partial(-P + \tau_{xx})}{\partial x} + \frac{\partial \tau_{yx}}{\partial y} + \frac{\partial \tau_{zx}}{\partial z} + S_x = \frac{\partial(\rho u)}{\partial t} + \nabla.(\rho u u) \tag{44}$$

Similarly, Total Y-Momentum

$$\frac{\partial(\tau_{xy})}{\partial x} + \frac{\partial(-P + \tau_{yy})}{\partial y} + \frac{\partial \tau_{zy}}{\partial z} + S_y = \frac{\partial(\rho v)}{\partial t} + \nabla.(\rho u v) \tag{45}$$

And for the Z-Momentum

$$\frac{\partial(\tau_{xz})}{\partial x} + \frac{\partial(\tau_{yz})}{\partial y} + \frac{\partial(-P + \tau_{zz})}{\partial z} + S_z = \frac{\partial(\rho w)}{\partial t} + \nabla.(\rho u w) \tag{46}$$

The sign of the Pressure term is opposite to the stress term in the same direction because, normally, the sign convention for the normal tensile stress is positive. But, since the pressure term is compressive, their signs are opposite.

2.4 Conservation of energy equation in three dimension

The energy equation is derived from the first law of thermodynamics, which states that the rate of increase of energy on a particle is equal to the sum of the net rate of heat addition on to the fluid particle and the work done on the particle.

Rate of Energy on a fluid particle

=

Net rate of heat added

+

Work done on a fluid particle

The rate of increase of Energy of fluid per unit volume is given as

$$\left[\frac{\partial(\rho E)}{\partial t} + \nabla.\rho E u\right] = \rho\frac{DE}{Dt} \tag{47}$$

2.4.1 Work done on a fluid particle

Work done by the surface forces per unit volume are equal to the stress and pressure terms multiplied by the velocity. The sum of these terms (work done) can be obtained multiplying the terms we derived by the Momentum equation with the velocity components [8].

Pressure terms:

$$-\left[\frac{\partial(pu)}{\partial x} + \frac{\partial(pv)}{\partial y} + \frac{\partial(pw)}{\partial z}\right] = -\nabla.(pu) \tag{48}$$

The work done due to the stresses is given as.

Total surface stress = Stress in X-direction + Stress in the Y Direction + Stress in the Z Direction

$$\left[\frac{\partial(u.\tau_{xx})}{\partial x} + \frac{\partial(u.\tau_{yx})}{\partial y} + \frac{\partial(u.\tau_{zx})}{\partial z}\right] + \left[\frac{\partial(v.\tau_{xy})}{\partial x} + \frac{\partial(v.\tau_{yy})}{\partial y} + \frac{\partial(v.\tau_{zy})}{\partial z}\right] + \left[\frac{\partial(w.\tau_{xz})}{\partial x} + \frac{\partial(w.\tau_{yz})}{\partial y} + \frac{\partial(w.\tau_{zz})}{\partial z}\right] \tag{49}$$

Therefore, the total work done on a fluid particle can be given as

$$-\nabla.(pu) + \left[\frac{\partial(u.\tau_{xx})}{\partial x} + \frac{\partial(u.\tau_{yx})}{\partial y} + \frac{\partial(u.\tau_{zx})}{\partial z}\right] + \left[\frac{\partial(v.\tau_{xy})}{\partial x} + \frac{\partial(v.\tau_{yy})}{\partial y} + \frac{\partial(v.\tau_{zy})}{\partial z}\right] + \left[\frac{\partial(w.\tau_{xz})}{\partial x} + \frac{\partial(w.\tau_{yz})}{\partial y} + \frac{\partial(w.\tau_{zz})}{\partial z}\right] \tag{50}$$

The first term is Energy Flux due to heat Conduction.

The second component that contributes to the rate of Energy addition is the rate of heat flux.

From **Figure** 7, we can see the relation

$$-\frac{\partial q_x}{\partial x} - \frac{\partial q_y}{\partial y} - \frac{\partial q_z}{\partial z} = -\nabla.q \tag{51}$$

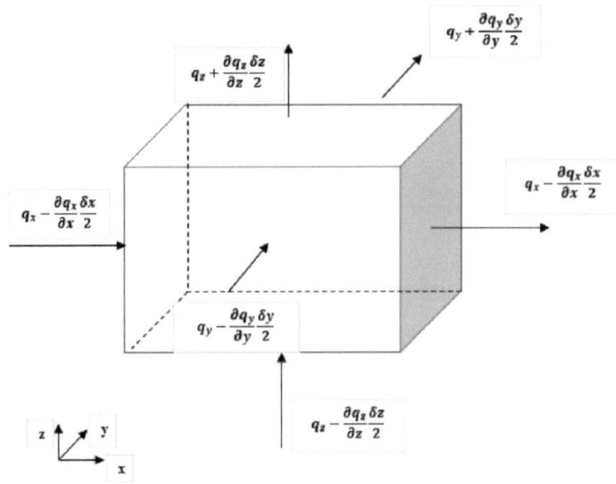

Figure 7.
Infinitesimal fluid element of heat energy.

But from Fourier's series, we can relate heat conduction to the temperature gradient as

$$q = -k\nabla T \tag{52}$$

Therefore, the energy addition due to heat is given as

$$div(k.grad.T) \tag{53}$$

2.4.2 Total energy equation

The energy equation mostly deals with the heat transfer analysis of a given system.

There are three main forms of energy namely, kinetic energy per unit mass 1/2 (u2 + v2 + w2), Internal (thermal) energy (i), and gravitational potential Energy [8]. The gravitational potential energy can be regarded as a source term as it does work when the fluid pass through a gravitational field.

As derived earlier, the energy equation is given as

$$\begin{aligned}\rho\frac{DE}{Dt} &= -\nabla.(pu) + \left[\frac{\partial(u.\tau_{xx})}{\partial x} + \frac{\partial(u.\tau_{yx})}{\partial y} + \frac{\partial(u.\tau_{zx})}{\partial z}\right] \\ &+ \left[\frac{\partial(\mathrm{v}.\tau_{xy})}{\partial x} + \frac{\partial(\mathrm{v}.\tau_{yy})}{\partial y} + \frac{\partial(\mathrm{v}.\tau_{zy})}{\partial z}\right] \\ &+ \left[\frac{\partial(\mathrm{w}.\tau_{xz})}{\partial x} + \frac{\partial(\mathrm{w}.\tau_{yz})}{\partial y} + \frac{\partial(\mathrm{w}.\tau_{zz})}{\partial z}\right] + div(k.grad.T) + \mathrm{SE}\end{aligned} \tag{54}$$

The mass and momentum conservation equations are solved for flow equations. Alongside them, energy equation is solved for heat transfer problems.

3. Navier–stokes equation

3.1 Basics of the Navier–stokes equation

The Navier- Stokes equation is a general equation that is used to understand properties of fluid flow. It is a core mathematical equation used to solve very important parameters like Pressure and velocity in the area of fluid Flow (fluid Engineering Simulation). It is basically the general conservation of momentum equation with the force (stress terms) more complicated and include viscosity terms as the viscosity induce stress on fluid particles.

One basic assumption here is that the fluid is considered isotropic. It does not behave differently at different points in nature.

In many fluid flows the viscous stresses can be expressed as functions of the local deformation rate or strain rate. In three-dimensional flows the local rate of deformation is composed of the linear deformation rate and the volumetric deformation rate.

And hence, the deformations can be found as

$$S_{xx} = \frac{\partial u}{\partial x}, S_{yy} = \frac{\partial v}{\partial y}, S_{zz} = \frac{\partial w}{\partial z} \tag{55}$$

These are the main stress terms.

For the linear shearing terms, which we have six of them, we have

$$S_{xy} = S_{yx} = \frac{1}{2}\left[\frac{\partial u}{\partial y} + \frac{\partial v}{\partial x}\right], S_{yz} = S_{zy} = \frac{1}{2}\left[\frac{\partial v}{\partial z} + \frac{\partial w}{\partial y}\right], S_{xz} = S_{zx} = \frac{1}{2}\left[\frac{\partial u}{\partial z} + \frac{\partial w}{\partial x}\right] \tag{56}$$

The volumetric deformation can be obtained by

$$\frac{\partial u}{\partial x} + \frac{\partial v}{\partial y} + \frac{\partial w}{\partial z} = div(u) \tag{57}$$

In fluid flow, there are two viscous term constants of proportionality, namely the linear viscous term (μ), and the volumetric viscous constant (λ). But the volumetric viscosity constant can be considered negligible.

$$\tau_{xx} = \left[2\mu\left[\frac{\partial u}{\partial x}\right]\right] + \lambda\nabla.u, \tau_{yy} = \left[2\mu\left[\frac{\partial v}{\partial y}\right]\right] + \lambda\nabla.u, \tau_{zz} = \left[2\mu\left[\frac{\partial w}{\partial z}\right]\right] + \lambda\nabla.u \tag{58}$$

And for the linear shearing viscous stresses, we obtain

$$\tau_{xy} = \tau_{yx} = \mu\left[\frac{\partial u}{\partial y} + \frac{\partial v}{\partial x}\right], \tau_{yz} = \tau_{zy} = \mu\left[\frac{\partial v}{\partial z} + \frac{\partial w}{\partial y}\right], \tau_{xz} = \tau_{zx} = \mu\left[\frac{\partial u}{\partial z} + \frac{\partial w}{\partial x}\right] \tag{59}$$

Inserting the general terms into the Navier stokes equation in the three dimensions yields,

$$\begin{aligned}
\rho\frac{Du}{Dt} &= -\frac{\partial p}{\partial x} + \left[\frac{\partial}{\partial x}\left[2\mu\frac{\partial u}{\partial x} + \lambda\nabla.u\right]\right] + \frac{\partial}{\partial y}\left[\mu\left[\frac{\partial u}{\partial y} + \frac{\partial v}{\partial x}\right]\right] + \frac{\partial}{\partial z}\left[\mu\left[\frac{\partial u}{\partial z} + \frac{\partial w}{\partial x}\right]\right] + S_x \\
\rho\frac{Dv}{Dt} &= -\frac{\partial p}{\partial y} + \frac{\partial}{\partial x}\left[\mu\left[\frac{\partial u}{\partial y} + \frac{\partial v}{\partial x}\right]\right] + \left[\frac{\partial}{\partial y}\left[2\mu\frac{\partial v}{\partial y} + \lambda\nabla.u\right]\right] + \frac{\partial}{\partial z}\left[\mu\left[\frac{\partial v}{\partial z} + \frac{\partial w}{\partial y}\right]\right] + S_y \\
\rho\frac{Dw}{Dt} &= -\frac{\partial p}{\partial z} + \frac{\partial}{\partial x}\left[\mu\left[\frac{\partial u}{\partial z} + \frac{\partial w}{\partial x}\right]\right] + \frac{\partial}{\partial y}\left[\mu\left[\frac{\partial w}{\partial y} + \frac{\partial v}{\partial z}\right]\right] + \left[\frac{\partial}{\partial z}\left[2\mu\frac{\partial w}{\partial z} + \lambda\nabla.u\right]\right] + S_z
\end{aligned} \tag{60}$$

Therefore, from this equation, we can rearrange the terms to obtain

$$\begin{aligned} &\frac{\partial}{\partial x}\left[\mu\frac{\partial u}{\partial x}\right]+\frac{\partial}{\partial y}\left[\mu\frac{\partial u}{\partial y}\right]+\frac{\partial}{\partial z}\left[\mu\frac{\partial u}{\partial z}\right] \\ &+ \\ &\frac{\partial}{\partial x}\left[\mu\frac{\partial u}{\partial x}\right]+\frac{\partial}{\partial y}\left[\mu\frac{\partial v}{\partial x}\right]+\frac{\partial}{\partial z}\left[\mu\frac{\partial w}{\partial x}\right]+\lambda\nabla.u=[S_x] \end{aligned} \tag{61}$$

Eq. (73) can be summarized as

$$= \nabla.(\mu\nabla U) + Sx \tag{62}$$

Therefore, the general Navier Stokes equations in three dimensions can be written as

$$\begin{aligned} \rho\frac{Du}{Dt} &= -\frac{\partial P}{dx}+\nabla.(\mu\nabla U)+S_x \\ \rho\frac{Dv}{Dt} &= -\frac{\partial P}{dy}+\nabla.(\mu\nabla V)+S_y \\ \rho\frac{Dw}{Dt} &= -\frac{\partial P}{dz}+\nabla.(\mu\nabla W)+S_z \end{aligned} \tag{63}$$

3.2 Basics of transport equation

The Transport equation traces the transport of a flow property Φ, which can be pollutants or temperature. The differential form of the Transport Equation can be written as [1, 8].

$$\frac{\partial(\rho\phi)}{\partial t}+\nabla.(\rho\phi u)=\nabla.(\Gamma\nabla\phi)+\mathrm{S_x} \tag{64}$$

The first term, $\frac{\partial(\rho\phi)}{\partial t}$ mentions the rate of increase of Φ in a fluid element. The second term is a convective term that represents convective outflow transport. The third term to the right of the equality sign is a diffusive transport term to mean rate of increase of ϕ due to diffusion.

The last term is the increase of ϕ in a fluid element due to production from the source.

3.3 Vorticity stream function

The vorticity stream function is a method of reducing unknowns and solving the Navier–Stokes equation.

Generally, for a two dimensional flow, we usually use three equations, two momentum equations in X and Y directions and one is Conservation of mass (Continuity equation) for case of incompressible flow [13].

$$\frac{\partial u}{\partial x}+\frac{\partial v}{\partial y}=0 \tag{65}$$

$$\frac{du}{dt}+u\frac{\partial u}{\partial x}+v\frac{\partial u}{\partial y}=-\frac{1}{\rho}\frac{\partial P}{dx}+\nu\frac{\partial^2 u}{\partial x^2}+\nu\frac{\partial^2 u}{\partial y^2}+S_{mx} \tag{66}$$

$$\frac{dv}{dt} + u\frac{\partial v}{\partial x} + v\frac{\partial v}{\partial y} = -\frac{1}{\rho}\frac{\partial P}{dy} + \nu\frac{\partial^2 v}{\partial x^2} + \nu\frac{\partial^2 v}{\partial y^2} + S_{my} \tag{67}$$

Here, in Eq. (63), we have three equations and three unknowns namely u,v and P. But, we can reduce the number of equations and the number of unknowns.

To do that, we can introduce a vorticity term.

$$\begin{aligned} \omega &= \nabla \times U \\ U &= f_n(u, v, w, t) \end{aligned} \tag{68}$$

The vorticity term has three components,

$$\omega_x = \frac{\partial w}{\partial y} - \frac{\partial v}{\partial z}, \omega_y = -\left[\frac{\partial u}{\partial z} - \frac{\partial w}{\partial x}\right], \text{and}\omega_z = \frac{\partial v}{\partial x} - \frac{\partial u}{\partial y} \tag{69}$$

Here, ω_x and ω_y contain terms varying with respect to Z, which we are not considering. Therefore, we shall take ω_z since it is valid and have variations X and Y, and not Z.

Therefore, differentiating the whole Y momentum equation with respect to X and the X momentum equation with respect to Y,

$$\frac{\partial(\mathrm{x} - momentum)}{\partial y} - \frac{\partial(\mathrm{y} - momentum)}{\partial x} \tag{70}$$

and finally subtracting it yields,

$$\frac{dw}{dt} + u\frac{\partial w}{\partial x} + v\frac{\partial w}{\partial y} = \nu\frac{\partial^2 w}{\partial x^2} + \nu\frac{\partial^2 w}{\partial y^2} + S_{mx} \tag{71}$$

Now let us introduce stream function to reduce the number of equations and unknowns. The Stream function combines velocity components u and v into one variable ψ. Let

$$u = \frac{\partial \psi}{\partial y} \text{ and } v = -\frac{\partial \psi}{\partial x} \tag{72}$$

Substituting Eq. (72) into (71), we obtain a single equation

$$\frac{\partial(\nabla^3\psi)}{\partial t} + \frac{\partial \psi}{\partial y}\frac{\partial(\nabla^2\psi)}{\partial x} - \frac{\partial \psi}{\partial x}\frac{\partial(\nabla^2\psi)}{\partial y} = \nu\left[\nabla^4\psi\right] \tag{73}$$

Which is a final vorticity stream function.

3.4 Classification of simple partial differential equations

Independent variable in a PDE can be either Temporal or Spatial, or only spatial in more than one dimension [8].

Let a PDE be of form

$$A\frac{\partial^2 \phi}{\partial x^2} + B\frac{\partial^2 \phi}{\partial x \partial y} + C\frac{\partial^2 \phi}{\partial y^2} + D\frac{\partial \phi}{\partial x} + E\frac{\partial \phi}{\partial x} + F\phi + G = 0 \tag{74}$$

Let coefficients be a,b,c,d,e,f and g be constants to maintain Linearity to make it simple.

The behavior of PDEs can be determined from its higher derivative terms, as in this case, the second order derivatives (**Table 2**).

It can be seen that the discriminant of the higher order terms are more determinant factor as to how a physical phenomenon modeled by a PDE behave [8].

Reduced form of Transport, Advection equation and Navier Stokes equations can be put in a form of a Matrix. Determinants and Eigenvalues can then be obtained from the Matrix, which can inform about the type of equation (PDE).

$$\mathrm{Det}|A - \lambda I| = 0 \tag{75}$$

If $\lambda = 0$, the equation is Parabolic,
$\lambda \neq 0$ and all are of the same sign, Elliptic,
$\lambda \neq 0$ and all but one are of the same sign, Hyperbolic.

3.5 Classification of fluid flow

Different flow types can be categorized, based on their properties, into different PDEs. Some of the flow types and their corresponding equation types are defined below.

As can be seen from **Table 3**, type of equation of Inviscid flow is different from that of the Navier Stokes equation because of the absence of a higher order term (Viscosity term).

Mach number is considered the measure of Compressibility of a fluid. It is the ratio of Speed of the Fluid to that of Speed of sound.

Flow with Mach number greater than one is termed as Supersonic flow and subsonic flow if less than one.

b^2-4 ac	Equation Type	Characteristics
>0	Hyperbolic	Two Real Characteristics
<0	Elliptic	No Real Characteristics
=0	Parabolic	One Real Characteristic

Table 2.
Equation types and their real characterstics.

Flow Type	Steady Flow	Unsteady Flow
Viscous Flow	Elliptic	Parabolic
Thin Shear Layer Flow	Parabolic	Parabolic
Inviscid Flow	M > 1, Hyperbolic M < 1,Elliptic	Hyperbolic
Navier Stokes Equation	Elliptic	Parabolic
Energy Equation	Elliptic	Parabolic

Table 3.
Flow types and their corresponding equations.

4. Conclusion

Fundamental derivation and discussions of Fluid dynamics is discussed in this chapter.

Initially, Mathematical background of fluid to help the reader sprint toward the area of fluids is discussed.

Green's theorem, Stoke 's theorem and Divergence theorem, which are basic tools to manipulate and convert among Line, surface and Volume integral was discussed briefly. The Divergence theorem is widely applicable to convert flux terms (Volume to surface integral) is discussed.

Leibniz's Rule, which helps solve Integro-differential equation was also included in this section.

Consequently, background on fluids was highlighted, followed by basic concepts that helps to understand fluids fundamentally, including appropriate scale of study, frame of reference and basic flow types.

In the main part of this chapter, Reynold's transport equation, which helps customize Lagrangian physical laws to Eulerian frame of reference is discussed.

Conservation of Mass and Momentum, which are fundamentals of fluid dynamics are discussed.

Energy equation, which is used to study heat flow within fluids are then discussed.

Considering Incompressible viscous flow, the famous Navier–stokes equation is then derived and discussed.

Application of Vorticity stream function, which is mostly used to solve the Navier–Stokes equation by reducing variables was derived and discussed.

Consequently, classification and characteristics of Partial Differential equations namely Parabolic, Elliptic and Hyperbolic equations are discussed.

Finally, classification of flows and their corresponding types like the Navier-Stokes equation, Inviscid and viscous flows, Compressible and Incompressible flows are then discussed.

Acknowledgements

I would like to thank my Jesus for his guidance in every aspect and way of my life. He has always been a father to me.

I would then like to extend my gratitude to Prof. Okey Oseloka Onyejekwe, for introducing me to Computational Fluid Dynamics, and the way to proceed.

I would like to my loving and beautiful wife Yordanos Kassa, who has always supported me and been on my side.

Conflict of interest

The author has no conflicts of interest to declare and there is no financial interest to report.

Author details

Nahom Alemseged Worku
Ethiopian Public Health Institute, Addis Ababa, Ethiopia

*Address all correspondence to: nahomal80@gmail.com

References

[1] The Finite Volume Method in Computational Fluid Dynamics An Advanced Introduction with OpenFOAM® and Matlab by F. Moukalled et.al.

[2] Advanced Engineering Mathematics, Second Edition by M.D. Greeenberg

[3] Morse PM, Feshbach H (1953) Methods of theoretical physics, Part I. McGraw-Hill, New York

[4] Fletcher, C. A. J. (1991). Computational Techniques for Fluid Dynamics, Vols I and II, Springer-Verlag, Berlin

[5] Stewart J (2008) Vector calculus, Calculus: early transcendental. Thomson Brooks/Cole, Connecticut

[6] Finite Element and Finite Volume Method for Simulation of Free Surface Flows: Application to Spillways by Fa tiha Lebdiri, Abdelghani Seghir and Ali Berreksi

[7] Numerical Calculations for a Boundary Layer Flow past a Moving Vertical Porous Plate with Suction/ Injection and Thermal Radiation by O. O.Onyejekwe, Yohannes Demiss Belete and Nahom Alemseged Worku

[8] An Introduction to Computational Fluid Dynamics THE FINITE VOLUME METHOD Second Edition H K Versteeg and W Malalasekera

[9] Finite Volume Methods and their Analysis by K.W.MORTON, E. SŪLI

[10] Finite Volume Method by S. Chakra verty, Et al

[11] Katz VJ (1979) the history of stokes's theorem. Math Mag (Math Assoc Am) 52:146–156 28. Morse PM, Feshbach H (1953) Methods of theoretical physics, Part I. McGraw-Hill, New York

[12] NUMERICAL AND NONLINEAR DYNAMICAL SYSTEMS ANALYSIS FOR FLUID MECHANICS AND HEAT TRANSFER PROBLEMS: Thesis by Nahom Alemseged Worku.

[13] Boundary Layer Theory by Herman Schlitching.

Chapter 2

Fluid Instabilities and Transition to Turbulence

Michael S. Roberts

Abstract

Fluid instabilities show up in many places in everyday life, nature and engineering applications. An often seemingly stable system with a gradient will often give rise to the development of instability, which can cascade eventually into turbulence. Governed by the parameters of the flow and fluids, when exposed to perturbation in the system, some wavelengths will grow, while others will not. This selectivity of specific structure sizes can be determined by using linear stability theory and then accounting for viscosity. Once these unstable wavelengths have grown to a substantial degree, the system typically becomes nonlinear before turbulence eventually sets in. Initially, looking at buoyancy-driven instabilities, one can clearly see how certain wavelengths can be selected. This can be extended to shear-driven instabilities and to geophysical systems. For some flows, simplifications can be made to analyze the specific fluid structures, while for others, only broad conclusions can be drawn about the stability criteria. With parallel shear flows (like that over wings and through pipes), the applications are more obvious, but the equations more difficult. However, conclusions can be drawn as to how one can control, prevent and initiate instability to suit our engineering needs.

Keywords: instability, turbulence, transition, Rayleigh-Taylor, Richtmyer-Meshkov, Kelvin-Helmholtz, Orr-Sommerfeld

1. Introduction

Fluid instabilities show up everywhere in nature. Fluid flow will start off laminar and smooth and then quickly transition to an irregular pattern eventually transitioning to turbulence. All you have to do to see its prevalence is look up at the sky on a cloudy day when the conditions are right such that there will be large clouds rolling past one another and spirals develop. This is the Kelvin-Helmholtz instability (where there is a shear between two fluids with different relative velocities). Another common instability is when you add cold milk to hot tea. The larger density of the cold fluid falls and displaces the hotter less dense fluid, and it is clear that specific structures form. A similar phenomenon occurs for contained flows, such as in pipes, or as in unbounded flow, such as that over a wing or out of a faucet. In all of these cases, solving too simplified equations of motion would lead to the solution that our ignorant view of the world might expect, where the fluids retain their smooth laminar structure, but this does not happen. Instead a specific size and shape of structure forms, usually in a periodic fashion, and this structure grows until finally becoming more and more chaotic until the flow is not longer laminar, but has transitioned to turbulence. A normal mode analysis can often be performed (where the equations of motion are assumed to be in a form where different wavelengths can be tested), and through solving the equation, the growth rate of various wavelength is found. This will tell us what specific configuration is expected.

When engineering new devices that involve fluid flows, it is important to account for these instabilities. This can be to try and minimize the onset of turbulence (or speed it up) or to understand the different wavelengths that may form and possibly causes resonances or unwanted behavior.

2. Stratified fluid instabilities

From a fluid instability perspective, stratified fluid systems are in some ways the easiest to understand and visualize, so we will start there. It is often the case when there is a fluid system in which two fluids with different properties, an unstable configuration can be realized. Here we will mainly consider the case where there is a clear boundary between two fluids, but this concept can be extended to a continuous variation between fluids, but the diffusion effects would damp out any perturbation and instability, so would not be as pronounced.

We can start our discussion by taking a vorticity perspective to the instability growth. If we start in two-dimensions with the inviscid Navier-Stokes equation and add a background velocity base state to the perturbed equation (here we are considering small perturbations to the base state), some conclusions can be drawn although not a rigorous derivation.

$$\frac{\partial \vec{u}}{\partial t}+\vec{u}\cdot\nabla\vec{u}=-\frac{1}{\rho}\nabla\vec{P}\Rightarrow\frac{\partial u_x+U_x}{\partial t}+(u_x+U_x)\frac{\partial(u_x+U_x)}{\partial x}+\frac{\partial u_y}{\partial t}+u_y\frac{\partial u_y}{\partial y}=-\frac{1}{\rho}\nabla\vec{P} \quad (1)$$

In Eq. 1, where u and U are the velocities of the perturbation and base state, respectively, ρ is the density and P is the pressure. We will neglect products of small quantities and also subtract the equation for the background flow. This will leave us with an extra term representing the product of our base velocity and perturbed velocity. We will now take the curl of this equation to arrive at a representation of the vorticity.

$$\nabla\times\left\{\frac{\partial u_x+U_x}{\partial t}+(u_x+U_x)\frac{\partial(u_x+U_x)}{\partial x}+\frac{\partial u_y}{\partial t}+u_y\frac{\partial u_y}{\partial y}=-\frac{1}{\rho}\nabla\vec{P}\right\} \quad (2)$$

$$\Rightarrow\frac{D\vec{\omega}}{\partial t}+\nabla\times\left(U_x\frac{\partial u_x}{\partial x}\hat{i}\right)+\nabla\times\left(-\rho^{-1}\nabla\vec{P}\right) \quad (3)$$

The expansion of the extra vorticity production term from the base velocity gives a few terms that can be neglected for this discussion $\Rightarrow\left(\frac{\partial}{\partial x}-\frac{\partial}{\partial y}\right)U_x\frac{\partial u_x}{\partial x}\hat{i}\Rightarrow -\frac{\partial U_x}{\partial y}\frac{\partial u_x}{\partial x}$. We will consider second-order derivatives of the perturbed quantities small enough to be neglected. We have also neglected the nonlinear convective terms coming out of the vorticity equation for simplicity. This yields:

$$\frac{D\vec{\omega}}{Dt}=\frac{1}{\rho^2}\nabla\vec{\rho}^{2}\times\nabla\vec{P}+\frac{\partial U_X}{\partial y}\frac{u_x}{\partial x} \quad (4)$$

From this we see that a pressure gradient across a density gradient can create vorticity (this is a buoyancy-driven instability) and a velocity gradient can create vorticity as well (shear-driven instability). Both of these require a perturbation at the interface to develop.

2.1 Buoyancy-driven instabilities

Buoyancy-driven fluid instabilities occur in a stratified fluid system when the light fluid is accelerated into the heavier one often by means of a pressure gradient. One way to understand this form of instability is from the baroclinic torque present at the stratified, perturbed interface. This baroclinic torque is created from the misalignment of the pressure and density gradients at the perturbed interface. When in the unstable configuration, for a particular harmonic component of the initial perturbation, this torque between the two fluids will create vorticity. This vorticity will impose a velocity field that will tend to increase the misalignment of the gradient vectors, which in turn will create additional vorticity, leading to more misalignment. This is observed in Eq. 4, where if we neglect the velocity gradient in the base flow as we have not considered that here, an increase in vorticity will be realized if $\frac{1}{\vec{\rho}^{2}} \nabla \rho \times \nabla \vec{P}$, which means that for instability $\nabla \vec{P} \cdot \nabla \vec{\rho} < 0$.

Illustrated in **Figure 1**, it is observed that the two counter-rotating vortices with strength ω have velocity fields that sum at the peak and trough of the perturbed interface. In the stable configuration the vorticity, and thus the induced velocity field, will be in a direction that decreases the misalignment and therefore stabilizes the system. In order for the instability to develop, $\nabla P \cdot \nabla \rho < 0$. This pressure is increasing in the direction from the more dense to the less dense fluid.

Two specific buoyancy-driven instabilities are the Rayleigh-Taylor (characterized by a constant acceleration) and Richtmyer-Meshkov (characterized by an impulsive acceleration).

2.1.1 Rayleigh-Taylor instability

The Rayleigh-Taylor instability (RTI) is a buoyancy-driven instability where the acceleration is constant with respect to the fluid flow time. The most notable example of the Rayleigh-Taylor instability is when a heavy fluid lies atop a light one while in the presence of a downward acting gravitational field. This instability is displayed in **Figure 2** of the experimental images of Roberts [1]. Here, the initially light over heavy stable fluid configuration is made unstable by accelerating the

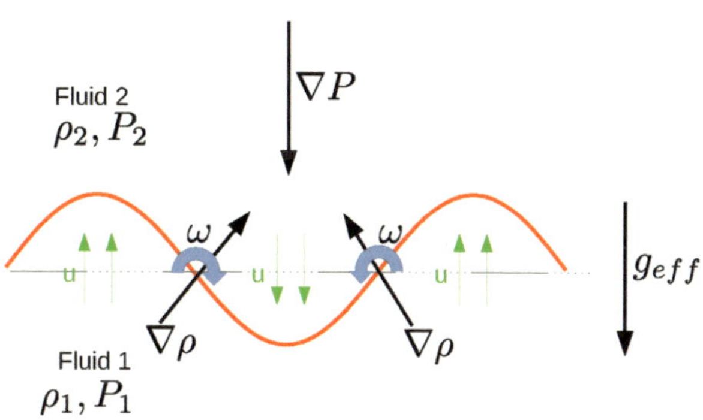

Figure 1.
Visualization of an unstable buoyancy instability configuration where baroclinic torque at the interface creates vorticity and induces a velocity field that increases the baroclinic torque.

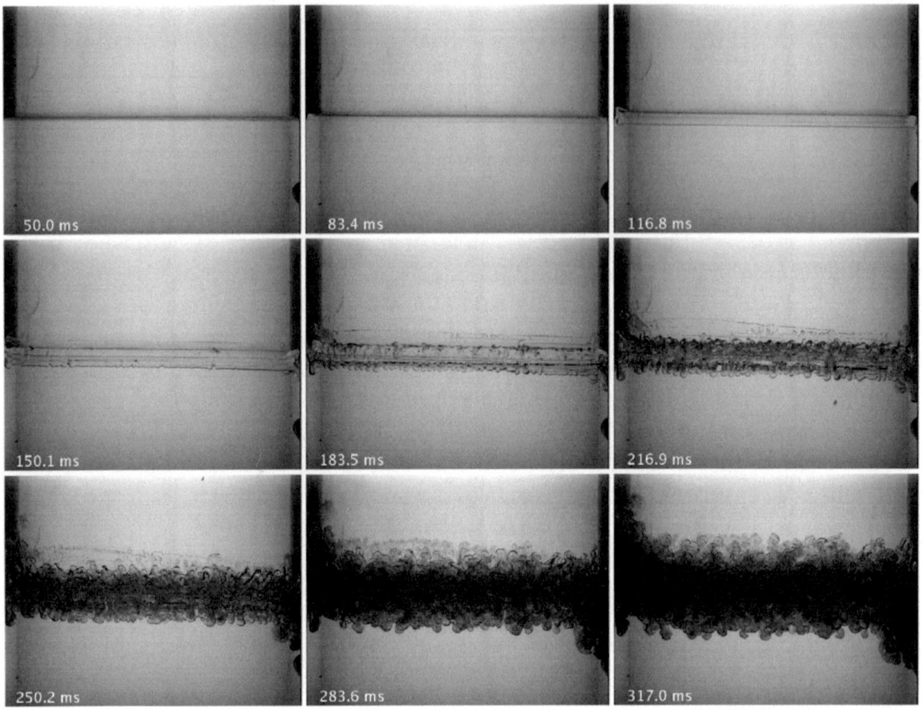

Figure 2.
Experimental images of Roberts [1] in which an unstable Rayleigh-Taylor configuration is formed where the light over heavy fluid system is made unstable. A progression of a specific wavelength is observed to develop and eventually a turbulent mixing region.

system downward at a rate greater than gravity (essentially switching the direction of gravity so that it is upward). As can be seen, a specific wavelength appears out of the background which grows and eventually creates a turbulent mixing region.

A simplified way to understand how we may have a Rayleigh-Taylor (RT) stable (or unstable) stratified configuration is by considering the situation in which there is an acceleration g_{eff} acting downward (in the negative z direction) on a stratified fluid system as depicted in **Figure 3** [1]. Considering a fluid particle, we can look at the forces acting on it in reference to the coordinate system in which z is directed upward. The acceleration produces a pressure gradient $\frac{\partial P}{\partial z} = -\rho g_{\text{eff}}$ inside the fluid which may create a force imbalance upon the fluid particle. If we choose a fluid particle in the upper fluid with density ρ_2, we see that the force, due to pressure, at the lower surface of this particle would be $[P_0 - \rho_1 g_{\text{eff}}\ell - \rho_2 g_{\text{eff}}(z - \ell)]A$ (where A is the area) and would be $[P_0 - \rho_1 g_{\text{eff}}\ell - \rho_2 g_{\text{eff}}(z - \ell + \Delta z)]A$ for the upper surface. We have chosen the geometry of the fluid particle here to simplify the equations. The force due to gravity on the fluid particle is $-\rho_2 \forall g_{\text{eff}} = -\rho_2 \Delta z A g_{\text{eff}}$ (where $\forall$ is the volume). Writing out Newton's second law we have (lower pressure force − upper pressure force + gravity force = mass × acceleration),

$$F = m\ddot{z} = \rho_2 g_{\text{eff}} \Delta z A - \rho_2 g_{\text{eff}} \Delta z A = (\rho_2 \forall + \rho_1 \beta)\frac{d^2 z}{dt^2} = 0, \tag{5}$$

where we have also included the added mass $\rho_1 \beta$ to account for the other fluid that must be accelerated away with the fluid particle. For this configuration, the fluid particle does not move, which is expected. If we interchange the fluid particle

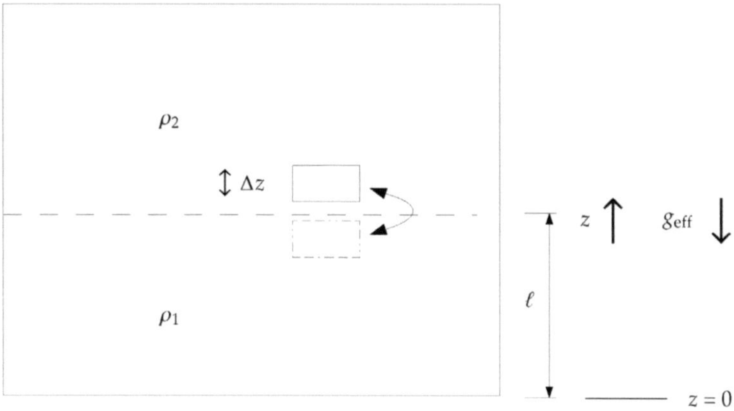

Figure 3.
A fluid particle in the upper fluid is interchanged with one from the lower fluid in a stratified system with downward acting acceleration [1]. Once displaced to the bottom fluid, the force balance on the particle might yield a configuration where it will continue to move from equilibrium. If $\rho_2 > \rho_1$, the fluid particle is accelerated further downward and the system is unstable. If $\rho_2 < \rho_1$, the fluid particle is pushed back across the interface and the system is stable.

with one from the lower fluid where the density is ρ_1, from Newton's second law, for the initial particle, we obtain (noting that the pressure forces have changed since we are in the lower fluid),

$$\begin{aligned}&\{[P_0 - \rho_1 g_{\text{eff}} z]A - [P_0 - \rho_1 g_{\text{eff}}(z + \Delta z)]A\} - \rho_2 g_{\text{eff}} \Delta z A \\ &= \rho_1 g_{\text{eff}} \Delta z A - \rho_2 g_{\text{eff}} \Delta z A = (\rho_2 V\!\!\!/ + \rho_1 \beta)\frac{d^2 z}{dt^2}.\end{aligned} \tag{6}$$

From this, if $\rho_1 > \rho_2$, the fluid particle is pushed back to where it came from (the system is stable). However, if $\rho_2 > \rho_1$, the fluid particle is pushed further away from where it originated and the system is unstable.

This concept of a fluid particle moving across the interface resulting in instability can be extended to the deflection of an interface in the Rayleigh-Taylor instability and illustrates the necessity of an initial perturbation on the interface since there is no mechanism to interchange a fluid particle across the interface. An example of a simple interface is shown in **Figure 4**, where the coordinate system is the same as in **Figure 3**. The interface has been deformed, simulating perturbations on the interface. For simplicity the geometry of the interface deformation has been chosen to be rectangular (the derivation here can be generalized to an individual Fourier mode so that any interface deformation would follow the same behavior). The fluid particle relocation is caused by deformation of the interface. The pressure force on the fluid particle's lower surface is $[P_0 - \rho_1 g_{\text{eff}} \ell]A$ and is $[P_0 - \rho_1 g_{\text{eff}} \ell - \rho_2 g_{\text{eff}} \Delta z]A$ for the upper surface. The force due to the weight of the fluid particle (which has density ρ_1) is $-\rho_1 V\!\!\!/ g_{\text{eff}} = -\rho_1 \Delta z A g_{\text{eff}}$. Note that once again we have chosen the interface deformation shape to simplify the calculations. We can then form the equation for the force balance as,

$$\rho_2 g_{\text{eff}} \Delta z A - \rho_1 g_{\text{eff}} \Delta z A = (\rho_1 V\!\!\!/ + \rho_2 \beta)\frac{d^2 z}{dt^2}. \tag{7}$$

In this arrangement, if $\rho_1 > \rho_2$, the fluid particle is pushed back to its original position (and thus the interface is brought back to equilibrium, so the system is

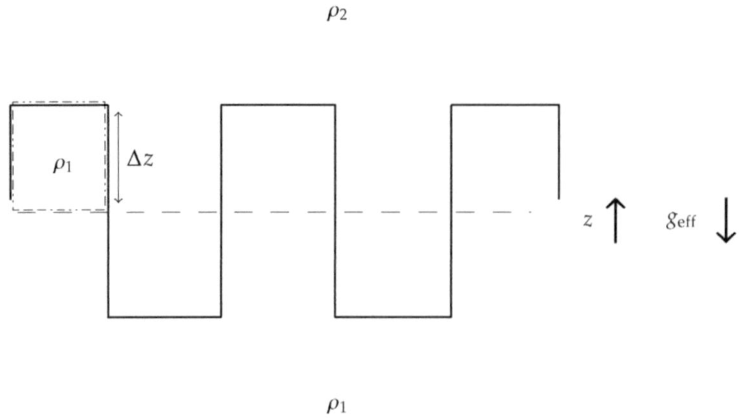

Figure 4.
Here an interface is shown [1] downward acting acceleration of a fluid particle displaced from the lower to upper fluid by means of interface deformation. If $\rho_2 > \rho_1$, the system is unstable (the fluid particle moves up farther from the center further deforming the interface). If $\rho_2 < \rho_1$, the fluid particle is moved back toward the center and the system is stabilized.

stable). However, if $\rho_2 > \rho_1$, the fluid particle is pushed further away from where it originated (deforming the interface further) and thus the system is unstable.

From this derivation, it is seen how the instability progresses, but does not say much about the initial stages (for that we need to use linear stability theory, Section 2.1.4) or late time. For the late time development, from Eq. (7), we can make some back of the envelope assumptions and arrive at a well known expression for the late time turbulent Rayleigh-Taylor instability. If we assume β in the added mass is the same as the Volume and rearrange and integrate twice, we can arrive at the well known expression:

$$h = \alpha A g t^2, \tag{8}$$

where α is the growth constant and A is the Atwood number (derived from the ratio of density difference to sum). This equation has been consistently found to fit the Rayleigh-Taylor instability in late time after it has become turbulent [2].

2.1.2 Richtmyer-Meshkov instability

We can extend our understanding of the Rayleigh-Taylor instability to that of the Richtmyer-Meshkov instability (RMI). From the vorticity argument for the instability it is obvious that all that is needed is $\nabla P \cdot \nabla \rho < 0$. The pressure term does not necessarily need to be constant as is for gravity, it can be impulsive as well. That is the case of the Richtmyer–Meshkov instability. In an instant, a large amount of vorticity gets deposited on the interface and the instability grows. The progression of the instability follows the progression of the constant acceleration case.

2.1.3 Transition to turbulence

The evolution of the Rayleigh-Taylor instability follows four main stages. Initially, if the perturbation amplitudes are small when compared to wavelength, the growth is exponential (following linear stability theory). Eventually, this will form spikes (fluid structures of heavy fluid growing into light fluid) and bubbles (fluid structures of light fluid growing into heavy fluid) from the individual sinusoidal modes on the interface. The growth of these structures can be modeled by using a buoyancy drag model and the growth is linear in time (the velocity is constant); this is the second stage [3]. At this time, non-linear terms in the equations of motion can no longer be ignored and mode-coupling will begin to play a role. Then, the spikes and bubbles interact with each other through bubble merging and competition, where fluid structures merge to create larger structures and larger structures envelop smaller ones respectively; this is the third stage. This eventually develops into a region of turbulent mixing, which is the fourth and final stage.

The mixing region that develops is believed to be self-similar and turbulent if the Reynolds number is large enough [4]. **Figure 5** represents the evolution of the Rayleigh-Taylor instability from small wavelength perturbations at the interface.

The turbulent mixing that takes place represents active-scalar, level 2 mixing where the mixing is coupled to the flow dynamics [4]. The flow is postulated to follow the model $h = \alpha A g t^2$, where h is the mixing layer width, $A \equiv \frac{\rho_2 - \rho_1}{\rho_2 + \rho_1}$ (the density contrast) is the Atwood number, g is the acceleration and t is time [5]. Under the self-similar hypothesis, the flow at different times has the same geometry and there is no obvious temporally constant length scale for the mixing region to be scaled with; the mixing layer width is only coupled to the length scales within the mixing region. Thus, the mixing layer width and the internal wavelengths increase in time and must grow proportionally with each other. Eventually, the range of scales within the mixing region form a sufficient inertial range for fully developed turbulence to be assumed. A derivation, through dimensional analysis, of this self-similarity is presented by Roberts [1]. A fully developed turbulent flow implies self-similarity, but since a self-similar flow does not necessarily imply turbulence,

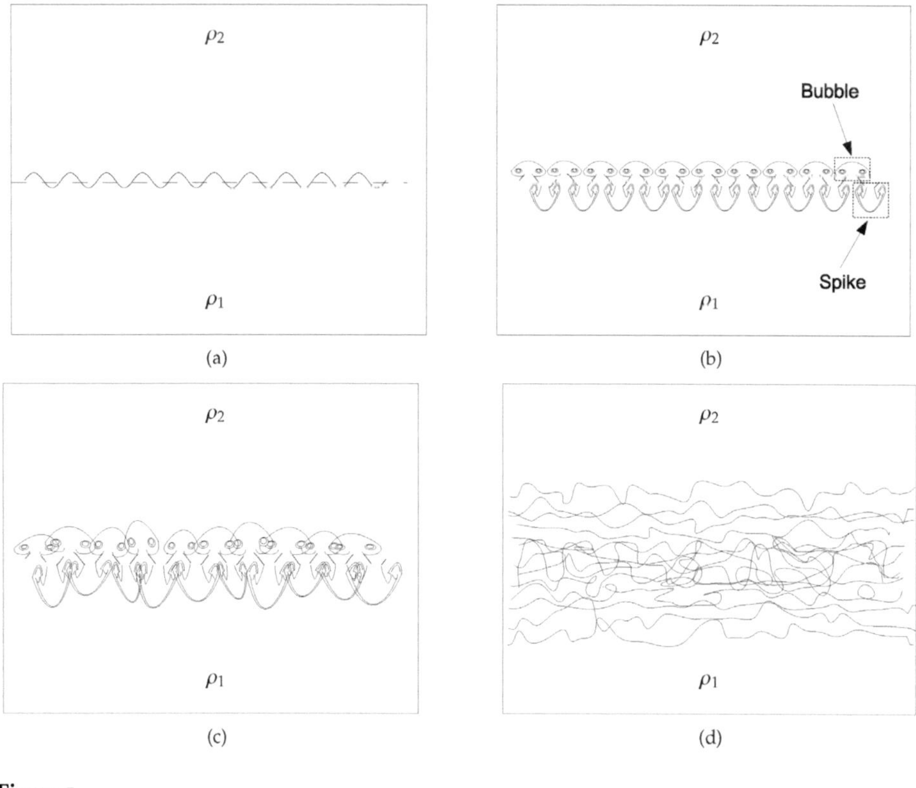

Figure 5.
This figure represents the evolution of the Rayleigh-Taylor instability from small wavelength perturbations at the interface (a) which grow into the ubiquitous mushroom shaped spikes (fluid structures of heavy into light fluid) and bubbles (fluid structures of light into heavy fluid) (b) and these fluid structures interact due to bubble merging and competition (c) eventually developing into a mixing region (d) [1].

turbulence cannot be assumed without quantifying the statistical properties of the flow or by making comparisons to other studies where flow statistics are quantified.

When studying the mixing region produced by the Rayleigh-Taylor instability, discrepancies between experiments and simulations make it obvious that it is necessary to verify that fully developed turbulence is indeed being reached. One method of doing this is to look at the spectra and verify that it obeys the Kolmogorov −5/3 energy cascade in the inertial subrange. Also, with the loss of initial conditions that is indicative of turbulent flow, verifying the self-preserving behavior of the flow is a possibility.

2.1.3.1 Kolmogorov energy cascade

It is well accepted that fully developed turbulence displays the $k^{-5/3}$ dependence for the velocity spectrum. The same wavenumber dependence will be present in an initially smooth scalar field that is disturbed by the same turbulence [6], which is often more testable for the buoyancy-driven instabilities we have discussed. One method of testing this is by using a normalized power spectrum of the FFT of the concentration profiles and comparing to that of the Kolmogorov −5/3 law. It was observed by Dalziel et al. [7] that the RT instability roughly fits this and by Ramaprabhu and Andrews [8] as well.

2.1.3.2 Self-preservation

An important aspect of fully developed turbulence is the concept of self-preservation. In the case of the turbulent Rayleigh-Taylor instability this would require the various turbulent properties along the mixing zone to have a shape that maintains itself in time. When normalized by the proper scale, the curves should collapse on top of each other.

Comparison of self-similarity is a difficult task when it comes to experiment, but Ramaprabhu and Andrews [8] does this with the use of PIV measurements. It is indeed observed that when normalizing with mean velocity, there is a collapse of the curves for profiles in later time. This can also be observed by looking at the similarity of different concentration profiles in time (**Figure 6**) from the experiments of Roberts and Jacobs [2]. From the profile images the self-similarity becomes obvious, thus implying turbulence.

2.1.4 Linear stability theory

Linear stability theory is often used derive equations governing the stability of a fluid system. It has been done many times in the past for the Rayleigh-Taylor instability. One such derivation is that done by Roberts which combines both the Rayleigh-Taylor and Richtmyer-Meshkov instabilities [9]. The way in which the derivation begins, is by considering a slightly perturbed interface and plugging this into the Navier-Stokes equations. By assuming small perturbations of the interface, simplifications can be made since squares of small values should be neglected. The full derivation will not be performed here, just some notable points for discussion.

We consider two stratified incompressible fluids where the interface is assumed infinitesimally thin and a sinusoidal disturbance is imposed upon it, in both the x and y directions, as displayed in **Figure** 7. Since each fluid region is considered to be initially at rest, they are irrotational. Here vorticity can only be introduced at the boundaries (in this case the interface) and then transmitted into the rest of the flow by viscous diffusion.

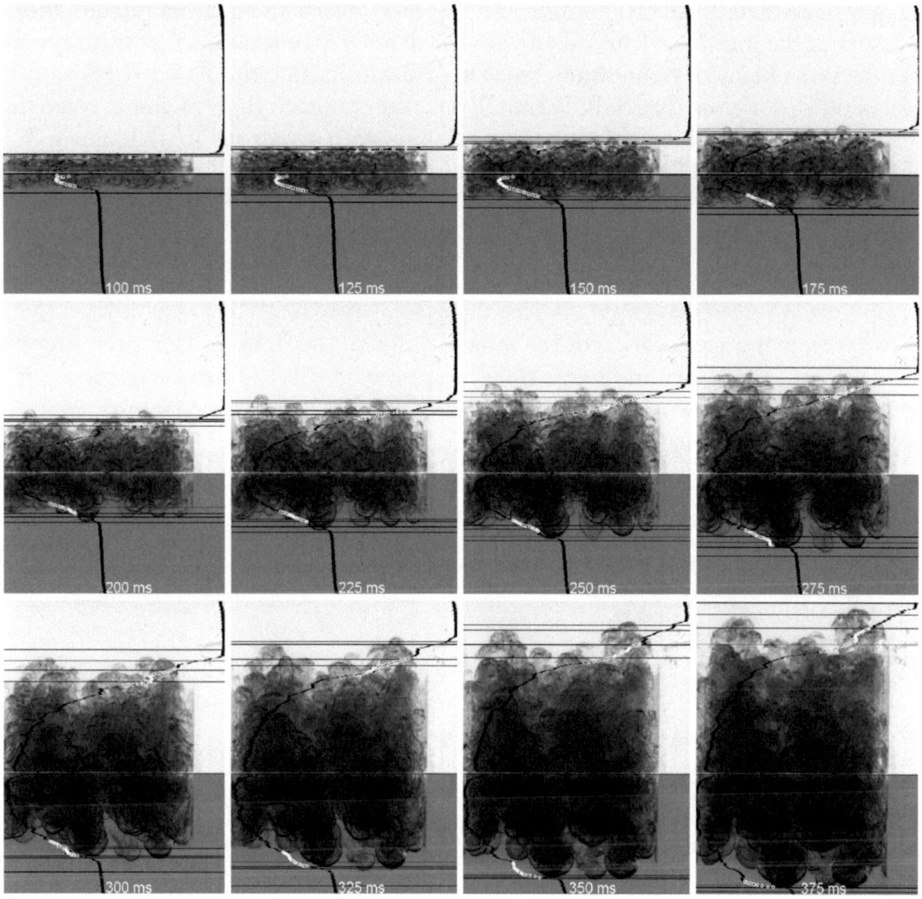

Figure 6.
From the experiments of Roberts and Jacobs [2], an experimental sequence of images where the images of an ensemble average of many experiments is shown progressing in time. Horizontally averaged intensity values are superimposed on the images. The profiles have the characteristics of a self similar flow as time progresses.

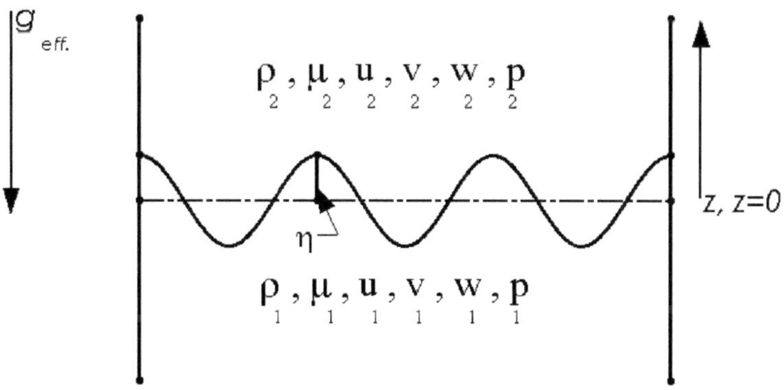

Figure 7.
Interface representing our fluid configuration [1].

Considering diffusion effects to be confined to the infinitesimally thin interface we can say that the fluid is irrotational throughout the two regions. By using the potential functions: $u = \frac{\partial \phi}{\partial x}$, $v = \frac{\partial \phi}{\partial y}$ and $w = \frac{\partial \phi}{\partial z}$ (where ϕ is a function of x, y, z and t), one can then derive the continuity equations for the two fluid regions. In general on the interface there is surface tension and this can give us an equation relating the pressure at the interface. This taken with the geometry of the interface itself gives an interfacial boundary condition. Since we are considering the flow to be irrotational, we can use the unsteady Bernoulli equation for each fluid region to solve the equations. After much equation manipulation one can arrive at the well known linearized stability ODE,

$$\ddot{a} - \left[\frac{\rho_2 - \rho_1}{\rho_2 + \rho_1} - \frac{\gamma k^2}{(\rho_2 + \rho_1) g_{\text{eff.}}}\right] k g_{\text{eff.}} a = 0, \tag{9}$$

where a is the acceleration of the interface due to the fluid flow, $\rho_{2,1}$ represents the density of the upper and lower fluids respectively, γ is the surface tension, $g_{eff.}$ is the effective gravity and k is the wavenumber (2π over the wavelength). By neglecting surface tension, one can simplify the expression to,

$$\ddot{a} - Akg_{\text{eff.}} a = 0. \tag{10}$$

This equation applies both to the Rayleigh-Taylor and Richtmyer-Meshkov instabilities.

2.1.4.1 Rayleigh-Taylor instability

In Eq. (10), when considering a constant acceleration for $g_{eff.}$, we have an ODE that represents the R-T instability. This is a second-order ODE and can be solved easily.

Representing $\sigma^2 = kAg_{\text{eff.}}$ we arrive at

$$a = C_1 e^{\sigma t} + C_2 e^{-\sigma t}. \tag{11}$$

If $Ag_{\text{eff.}} < 0$, σ is imaginary and therefore the equation yields a stable condition.
If $Ag_{\text{eff.}} > 0$, σ is real and therefore the equation is unstable (it grows in time).

This equation can also be represented in a better way in which the coefficients represent the initial amplitude and velocity of the sinusoidal disturbance,

$$a = a_0 \cosh(\sigma t) + \dot{a}_0 \sinh(\sigma t). \tag{12}$$

2.1.4.2 Richtmyer-Meshkov instability

Eq. (10) can also be considered when $g_{eff.}$ is an impulsive acceleration defined by $g_{eff.} = \delta(t)V$, where V is the velocity produced by the impulse.

This can be integrated to yield

$$\int_{a_0^-}^{a_0^+} \ddot{a} = AkV \int_{t^-}^{t^+} a\delta(t)d\tau, \tag{13}$$

where

$$a_0 = a(0). \tag{14}$$

$$\Rightarrow \dot{a} = AkVa_0 + \dot{a}(0).$$

by integrating once more we acquire

$$a = AkVa_0t + \dot{a}_0t + a_0 \Rightarrow a = a_0(kAVt + 1) + \dot{a}_0t. \tag{15}$$

2.1.5 Small wavelength damping and stabilization

The effects of viscosity act only at small scales in RT instability and therefore act to select particular wavelengths as opposed to others. Since viscosity only acts at small scales, its effect can be neglected once the instability has become larger than these scales. This can be understood by comparing the terms of RT growth with that of viscous damping. First, The RT growth term from inviscid theory is $e^{\sqrt{kAg_{\text{eff}}}t}$ [9] and that of viscous damping is $e^{-2\nu k^2 t}$ [10]. It is of interest here to see when the RT growth term is much larger than the viscous term,

$$e^{\sqrt{kAg_{\text{eff}}}t} > > e^{-2k^2\nu t}. \tag{16}$$

This yields

$$k < < \sqrt[1/3]{\frac{Ag_{\text{eff}}}{4\nu^2}}. \tag{17}$$

Eq. (17) gives us a way to calculate an approximate wavelength at which larger than which we can ignore viscosity effects. As an example, with water and air at room temperature, $k = 13,485$. This translates to a wavelength of approximately 0.46 mm. The dominant scales that we are measuring are certainly larger than this, so for this regime we can neglect viscous effects. Next, for completeness the RT growth from the self-similar model will be compared to the viscous damping term. Again, let us examine when the RT growth is a lot larger than the viscous damping,

$$\alpha Ag_{\text{eff}}t^2 > > e^{-2k^2\nu t}. \tag{18}$$

This yields

$$k < < \sqrt{\frac{\ln\left(\alpha Ag_{\text{eff}}t^2\right)}{-2\nu t}}. \tag{19}$$

As an example, from the experiments of Roberts and Jacobs [2], by assuming an approximate α value of 0.05 and a time of 300 ms (this time corresponds to the beginning of the measurable mixing region development in their experiments) we conclude that the wavenumbers should be less than approximately 2500 (2.5 mm). This does fall in line with the small wavelengths and structures we observe in the images shown previously.

Keep in mind, this is an approximation to the viscous effects where a viscous damping was used. If a detailed analysis is necessary to determine the fastest growing wavelength, the derivations of Chandrasekhar [11] will give a more detailed explanation and more exact solution.

Another damping effect at small wavelengths is that due to interfacial tension. We use the term interfacial tension here to be more general, but the most obvious example of this is surface tension where there is a free surface - such as in the air water interface created in a glass of water or a straw. Interfaces of oil and water will also have an interfacial tension which will act to shift the fastest growing wavelength to the larger scales by damping out smaller scales as well as viscosity. However, interfacial tension has more than just a damping effect but actually has a stabilizing effect, such that there becomes a critical wavelength smaller than which the instability will not grow.

2.1.6 Applications

The classic observation of the Rayleigh-Taylor instability is the simple inversion of a glass of water. Due to gravity, air (the less dense fluid) moves into the water (the more dense fluid) and the instability develops as a consequence of perturbations at the interface (as typically is in a natural environment). If one were to look at simple fluid statics, the water should not "fall" out. Since the bottom of the glass is covered, atmospheric pressure should hold the water in place when fluid statics alone is considered; this does not take place because of the instability. An interesting phenomenon is that of a covered straw with water in it. The same configuration is present, however the water stays in place. By considering surface tension, we must recognize that it has a stabilizing affect on the instability for smaller wavelengths. The diameter of the straw is often smaller than the critical wavelength and therefore nothing larger can develop and the wavelengths that do are not unstable. This has implications for any engineering applications that rely on gravity. If the diameter is too small, you cannot rely on the presence of the instability to assist and you would need to account for the pressure difference given by fluid statics and atmospheric pressure. An extension of this example can be made to the Richtmyer-Meshkov instability. With the straw example, instead of just allowing the water to remain under gravity, you can shake the straw thus creating impulsive accelerations that (if strong enough) may shift the critical wavelength low enough such that the flow becomes unstable and the liquid flows out. The same can be thought of for the bottle of ketchup that needs that impulsive acceleration to start flowing. Although the physics of the ketchup are more complicated since it is a non-Newtonian fluid, at least at the beginning of the flow, RM instability plays a role. The extension here to our pipe under gravity example would be that in a situation where some external pressure differential is applied, but not necessarily enough to overcome atmospheric pressure and the diameter of the pipe is too small for the RT instability alone to work, pulsing the pressure source might help trigger a RM instability.

A more natural occurrence of these instabilities is in supernovae. Here, there are stratified gases of different density. This difference in density arises from the fact that the gas closer to the center is hotter (and therefore less dense), due to its proximity to the burning fuel, than the gas farther from the center. This, in addition to the outward acceleration that was produced by the explosion (both impulsive and constant), creates unstable RT and RM configurations [12]. This in turn will generate mixing which will alter the way in which the flow progresses and how heat is distributed. Astronomers can use this information to better understand and find these phenomena. In relation to studying the stars, RTI also shows up when we explore them. Hall-effect thrusters are becoming very popular for space flight (especially satellites) due to their large specific impulse. In these thrusters, the wall at the thruster exit has been shown to erode due to the instability and therefore a better understanding is necessary [13]. We also see RTI in salt domes. Here, the less dense salt that is buried beneath more dense sediment experiences an upward acceleration due to gravity [14]. Although the timescales and effective viscosities are very large, this still forms a RTI on geologic timescales.

Another important application is inertial confinement fusion (ICF), which if mastered would lead to cheap and plentiful energy from water. In ICF, a capsule containing a Deuterium/Tritium (DT) mixture is bombarded with energy originating from high powered lasers with the purpose of causing a fusion reaction to take place;

the two isotopes fuse producing He_4, a neutron and energy [15]. ICF experiments are currently being performed at the National Ignition Facility (NIF) in Lawrence Livermore National Laboratory (LLNL). The ICF capsule is a sphere comprised of three main layers. The outer shell is an ablator material made from plastic doped with other elements such as Beryllium or Germanium. Interior to that is a layer of DT ice surrounding DT gas. There are two main types of ICF, direct and indirect drive. In direct drive, lasers directly irradiate the target. In indirect drive, lasers enter a hohlraum which has the capsule in the center. The hohlraum is a hollow cylinder that is composed of a high Z (large atomic number) material, such as gold. The lasers irradiate the inside of the hohlraum which re-emits the energy as x-rays. In the indirect drive method, a more uniform energy distribution is deposited on the ablator layer. The energy deposited on the ablator causes it to blow off, and by Newton's third law, PdV work is done on the interior of the capsule. The compression of the DT gas region results in an increase in pressure at the center of the capsule causing very high temperatures to develop. In addition, shocks (caused by the ablation) pass into the DT gas region, which also add to the pressure and temperature rise. The pressure rise at the center eventually acts to decelerate the initially accelerating implosion until a stagnation point is reached [16]. This "hot spot" will reach the conditions for thermonuclear burn if a high enough temperature is achieved. During this process, there are two ways in which the Rayleigh-Taylor and Richtmyer-Meshkov instabilities can develop which acts to mitigate ignition and decrease total yield.

Firstly, RTI and RMI can occur at the interface of the outer ablator shell (after becoming a plasma) and the DT ice layer during the initial implosion of the target. In this configuration, the smaller density of the outer ablator plasma layer and the larger density DT ice inner layer create an inward acting density gradient. This in conjunction with the outward acting pressure gradient results in an RT/RM unstable configuration. By choosing layers of gradually varying density with different dopants such as Germanium, for the ablator material, the density difference can be decreased; thus, decreasing RT growth. Also, by using indirect drive (to produce a more uniform energy deposition), the effect of the instabilities can be minimized as well by effectively decreasing the perturbations necessary to begin the instability. The second way that RTI and RMI can occur is during the deceleration phase between the high temperature, high pressure DT gas and the outer, colder DT ice layer. Here, the pressure gradient is directed inward and the density gradient is directed outward which is also an RT/RM unstable configuration. The RTI generated in both these instances causes mixing. This mixing brings cold fuel from the outer layer into the center "hot spot," lowering the temperature and decreasing the reaction rate; this process may prevent ignition altogether [17]. By more fully understanding this instability, more efficient capsules can be designed. In addition to this, other methods to control the onset of RTI in fusion experiments using rotating magnetic fields is being studied [18].

There are some situations where one does not want to prevent these instabilities from forming at all, but actually want to encourage it and the increased mixing that happens from it. One such example for RMI is that for a scramjet. With scramjets (supersonic combustion ramjets), we wish to do combustion at supersonic speeds. This is as opposed to standard ramjets in which the flow is slowed in the engine so that proper mixing can occur. To accomplish this in scramjets, RMI is utilized to enhance the mixing [19]. The geometry of the engine can be configured to create shockwaves that will interact at specific fuel/air boundaries; this will impart impulsive accelerations that trigger RMI and eventually turbulent mixing. This extra mixing is necessary to get a proper fuel/air mixture.

As can be observed, there are many applications to studying these instabilities to control, prevent or encourage their growth.

2.2 Shear-driven instabilities

Another class of stratified instabilities is shear-driven ones—where there is a difference in shear forces across the interface for instance. One such instability is the Kelvin-Helmholtz instability that happens if there is a jump in velocity across an interface. From Eq. 4, we see that if the velocity gradient is large enough, then a stabilizing buoyancy instability effect will be overcome and we will have vorticity deposited. $\frac{1}{\rho^2}\nabla\vec{\rho}^{\,2} \times \nabla\vec{P} + \frac{\partial U_X}{\partial y}\frac{u_x}{\partial x} > 0$. As can be seen in **Figure 8**, when we ignore the effects of buoyancy, a velocity difference between top and bottom fluid (a gradient in the direction orthogonal to the flow) will create vorticity due to the torque from a perturbed interface. If a U_2 is larger than U_1, this vorticity will create even more shear, which will create more vorticity. This instability can be observed in clouds when there is stratification with high velocity present or even when one pours a bottle of oil and vinegar salad dressing. Another interesting application of this instability is in semiconductor manufacturing where the ion beams used in chemical vapor deposition and ion implantation becomes subjected to this instability [20].

3. Baroclinic instability

All of the instabilities mentioned so far often occur in nature as there are often stratified flows in the atmospheres and oceans. Another extension of this is on the much larger geophysical scale, where the Coriolis force due to the earth rotation and velocity difference at different lines of latitude as the radius with the rotation axis varies. In this configuration, we have both the hydrostatic balance $\frac{\partial p}{\partial z} = \rho g$ and the geostrophic balance $2\Omega sin\theta U = -\frac{1}{\rho}\frac{\partial p}{\partial y}$, where Ω is the earth's rotation and density ρ is a function of temperature (**Figure 9**).

It is then determined that for instability that, $\frac{H\sqrt{-\frac{g}{\rho_0}\left(\frac{\partial p}{\partial y}\right)}}{\Omega}$ [21], where H is the height in the vertical and $\frac{\partial p}{\partial y}$ is the pressure gradient due to a temperature gradient. As can be observed, there are very specific conditions for this instability to develop. If the temperature gradient is too small or too high, this particular instability will not develop, but also a part of this is the vertical height and the earth's rotation. This instability shows up quite often as the development of vorticity in the earth's oceans and atmosphere and is a large contributing factor to weather patterns.

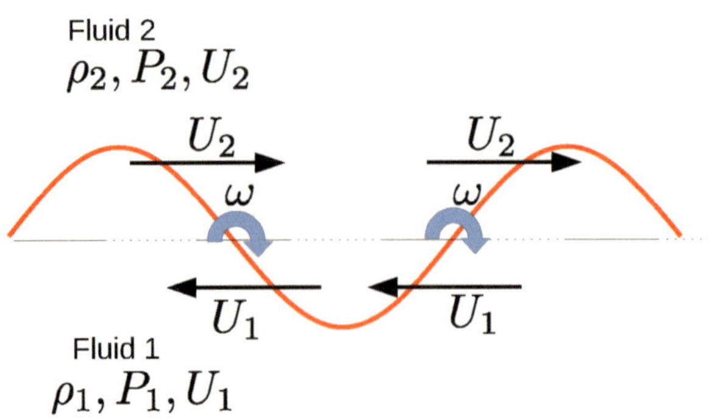

Figure 8.
Vorticity created when there is velocity gradient in the base flow and a perturbed interface. It is clear here that a torque would be created when the interface is misaligned.

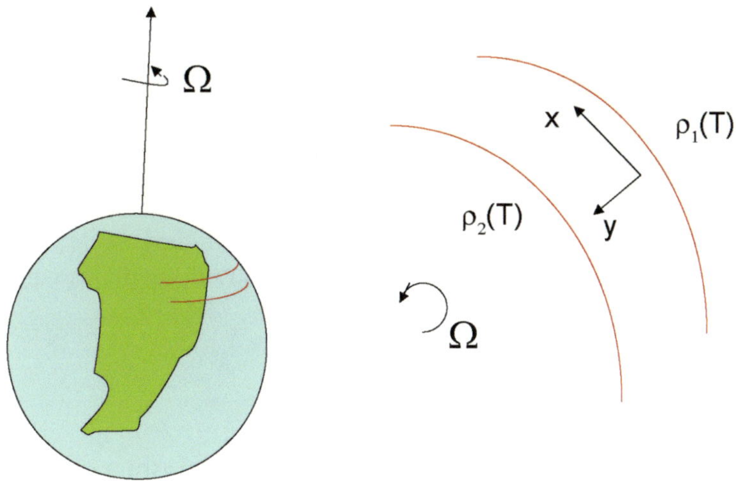

Figure 9.
The rotation of the earth sets up a situation in which instability can occur due to the combined action of the hydrostatic and geostrophic balances.

4. Parallel shear flow instability

A very broad and far reaching class of flows is that of parallel shear flows. These are flows that are mostly unidirectional and have velocity profiles created by the no slip condition at the boundary. This includes and has historically been studied for pipe flow by Reynolds [22]. The flow can also be extended to that of unbounded flows and even base flows in which there is a slight curvature, but parallel to a first-order approximation, such as flow over a wing. Under certain conditions, these flows can become unstable leading to turbulence and separation which has implications to engineering design. First, we will consider inviscid flow and then discuss viscosity effects.

4.1 Inflection points and adverse pressure gradients in inviscid flow

It was Rayleigh [23] who first recognized the importance of inflection points in the velocity profile. He proposed that for instability to be present, a necessary condition is an inflection point in the velocity profile. This was expanded upon by Fjørtoft [24] who proposed that another necessary condition for instability is that $\frac{\partial^2 U}{\partial z}(U - U_s) < 0$ somewhere in the flow where z_s is a point at which $\frac{\partial^2 U}{\partial z} = 0$ and $U_s = U(z_s)$ [25]. This basically means that in a velocity profile, not only does there need to be an inflection point (change in curvature or where it goes from concave up to down or vice versa), also, if you follow along the profile, at some points in the flow, the difference in velocity there to that at the inflection point times the curvature should be negative. This typically happens when there is an adverse pressure gradient (pressure that pushes in the opposite direction of the flow). In this situation, the velocity profile starts as one would expect (bulging forward), but it eventually starts bulging backwards which is unstable and can lead to flow separation in an unbounded flow.

One way that an inflection point occurs is with an adverse pressure gradient. In pipe/duct flow this can be difficult to realize if the main flow is caused by a pressure gradient. However, we can have a situation where a localized pressure gradient (caused by a fan or impeller) creates a forward moving velocity profile, but there is back pressure in the over all configuration that will make an unstable flow configuration. Adverse pressure gradients are a bit more obvious with unbounded flows. We can start by looking at the boundary layer equation,

$$u\frac{\partial u}{\partial x}+v\frac{\partial u}{\partial y}=-\frac{1}{\rho}\frac{\partial p}{\partial x}+\nu\frac{\partial^2 u}{\partial y^2} \tag{20}$$

where the pressure gradient is based has the form from the base flow as $-\rho U\left(\frac{dU}{dx}\right)$. In this situation, at the wall (due to the no slip condition and continuity equation), we have zero velocity components which yields,

$$\mu\left(\frac{\partial^2 u}{\partial y^2}\right)_{wall}=\frac{\partial p}{\partial x}. \tag{21}$$

Therefore, in the immediate vicinity of the wall, the curvature of the velocity profile is dictated by whether the pressure gradient is positive or negative. Thus, an adverse pressure gradient will lead to an inflection point. Over a wing, what creates lift (and a pressure gradient in the y direction) is streamline curvature. The streamlines must curve to fit the body at first, but this also means that toward the trailing edge they must then curve back to the background stream, and there will be an adverse pressure gradient. Decreasing the degree of curvature over the trailing edge will decrease the magnitude of the adverse pressure gradient and thus prevent separation further along the wing, thus minimizing drag. In addition, "tripping" the boundary layer by making it turbulent early on also prevents separation as there is more momentum in the flow giving a wider velocity profile that can withstand an adverse pressure gradient longer.

4.2 Orr-Sommerfeld equation

The criteria for stability so far have been considered for inviscid flow. To extend this to include viscosity as well does make the equations much more complicated. We will not derive this here. It can be derived by using linear stability theory on the Navier-Stokes equations as done by Cohen and Kundu [21] yielding the result,

$$(U-c)\left(\frac{\partial^2\phi}{\partial y^2}-k^2\phi\right)-\frac{\partial^2 U}{\partial y^2}\phi=\frac{1}{ikRe}\left[\frac{\partial^4\phi}{\partial y^4}-2k^2\frac{\partial^2\phi}{\partial y^2}+k^4\phi\right], \tag{22}$$

where c is the wave speed, k is the wavenumber and ϕ is defined such that the perturbation velocities, $u=\frac{\partial\phi}{\partial y}$ and $v=-ik\phi$. As can be observed, this is a fourth-order differential equation which is very difficult to solve. The only way to approach such a problem would be to solve this equation a numerical simulation. Since it is an ordinary differential equation, there are many methods that can be used. But, better yet is to use one of the inviscid simplifications discussed previously.

4.3 Engineering applications

Parallel shear flows have become some of the most obvious flows around us and thus have strong engineering importance. From pipe flow to flow over wings and cars there is great importance. With flow in pipes, it was noticed by Rayleigh [23] that turbulent spots develop above a certain Reynolds number and then eventually the flow becomes fully turbulent. This happens at a Reynolds number of approximately 3000. This is important to take note of as a turbulent flow will be noisier and

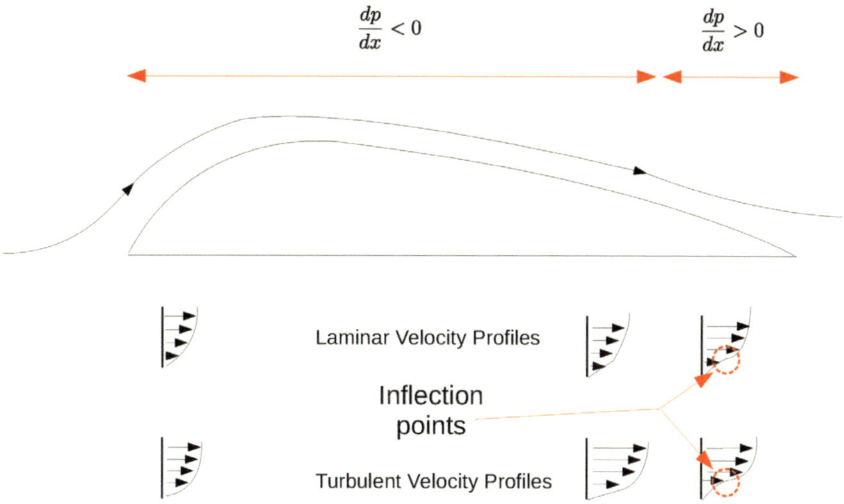

Figure 10.
Initially, the flow is in a favorable pressure gradient, but this eventually changes to adverse one where an inflection point forms in the velocity profile and the flow is in danger of separation which would increase drag. The top row of velocity profiles represent that for laminar flow, whereas the bottom row represents that for a turbulent flow which is seen to be fuller and thus less susceptible to backflow.

the extra flow non-uniformity can lessen the life of pipes. For flow over bodies, we have to also look at the pressure gradient and whether it is adverse or not. An adverse pressure gradient will eventually lead to separation, but this can be delayed if the flow is turbulent. This means, that although we need to locally introduce irregularities in the flow to induce turbulence, at the same time we would want to minimize the adverse pressure gradient to prevent separation (which will in turn increase drag). A pictorial representation of a simplified view of the flow over the wing is shown in **Figure 10**.

As can be observed from the flow over a wing, at first there is a favorable pressure gradient, which eventually turns into an adverse pressure gradient at the trailing edge. In order to make lift, we want to have a pretty large wing curvature at first (this creates the pressure gradient perpendicular to the wing that creates lift), but then we can smoothly allow the wing and streamlines transition to free stream to try and push back the section where we have an adverse pressure gradient. In this adverse pressure gradient region is where the velocity profile can get an inflection point which could eventually lead to back flow in the velocity profile and flow separation. Once the flow separates, you will have increased pressure drag as there will be a low pressure region in this separated trailing edge region. It also turns out that since turbulent boundary layers are fuller and have more momentum, it takes longer for an inflection point to create backflow and therefore the flow stays attached longer. One can "trip" the boundary layer (by depositing a small amount of vorticity in the flow right in the boundary layer (small triangular surfaces seen on a wing's leading edge) such that the boundary layer becomes turbulent, but the rest of the flow does not.

5. Conclusions

Fluid instabilities show up everywhere in nature. In this chapter we have discussed some of the main instability classes. Stratified fluid flows were discussed first as it is in some ways the simplest to understand. They show up in our coffee, our condiments, in the atmosphere and the oceans, basically wherever there is a density difference present between the two layers. Depending on the orientation of the density difference, vorticity can be generated in a self-sustaining process. Using linear stability theory, it is determined how different wavelengths grow at different rates for the system and how interfacial or surface tension acts has a stabilizing effect. One can also look at the effects of viscosity which has a damping effect on the smaller wavelengths, thus giving rise to a fastest growing wavelength which has direct implications to engineering applications as the geometry of the system can act to directly prevent or create these wavelengths, therefore controlling instability growth. Once the instability has grown enough, the linearized equations no longer hold and non-linear effects including interactions between structures will take place. This eventually leads to turbulence. This instability can then be extended to include the case where instead of a density difference at an interface, there is a velocity difference causing what is known as shear-driven instabilities. Once these two base instabilities are discussed, it is an easy extension to that of the baroclinic instability which is a primary cause of many of the earth's weather patterns. Due to interactions of velocity and density gradients (caused by temperature gradients), it is a natural extension to the previously discussed instabilities. From this, the more complicated case of parallel shear flows was discussed. This includes flow over wings and flows in pipes. In this case, first the stability criteria was discussed, but then extended to the full Orr-Sommerfeld equation, where unlike linear stability theory, does not remove viscosity from its initial derivations and allows for non-linear effects. Here we discuss instability more as an eventual path to turbulence and how controlling, preventing and even creating it can be advantageous to reducing drag, or preventing noise. In this chapter, much was covered but a broad understanding of how different fluid instabilities all relate together and can be understood to thus control them when designing and running our engineering systems.

Author details

Michael S. Roberts
University of Arizona, Tucson, AZ, USA

*Address all correspondence to: window59@hotmail.com

References

[1] Roberts MS. Experiments and simulations on the incompressible, Rayleigh-Taylor instability with small wavelength initial perturbation [Dissertation]. The University of Arizona; 2012

[2] Roberts MS, Jacobs JW. The effects of forced small-wavelength, finite-bandwidth initial perturbations and miscibility on the turbulent Rayleigh-Taylor instability. Journal of Fluid Mechanics. 2016;**787**:50-83

[3] Oron D, Arazi L, Kartoon D, Rikanati A, Alon U, Shvarts D. Dimensionality dependence of the Rayleigh-Taylor and Richtmyer-Meshkov instability late-time scaling laws. Physics of Plasmas. 2001;**8**(6):2883-2889

[4] Dimotakis PE. Turbulent mixing. Annual Review of Fluid Mechanics. 2005;**37**:329-356

[5] Youngs D. Numerical simulation of turbulent mixing by Rayleigh-Taylor instability. Physica D. 1984;**12**:32-44

[6] Tennekes H, Lumley JL. A First Course in Turbulence. Massachusetts, USA: MIT Press; 1972

[7] Dalziel SB, Linden PF, Youngs DL. Self-similarity and internal structure of turbulence induced by Rayleigh-Taylor instability. Journal of Fluid Mechanics. 1999;**399**:1-48

[8] Ramaprabhu P, Andrews MJ. Simultaneous measurements of velocity and density in buoyancy-driven mixing. Experiments in Fluids. 2003;**34**(1):98-106

[9] Roberts MS. Experiments on the miscible liquid Rayleigh Taylor and Richtmyer Meshkov instabilities [master's thesis]. University of Arizona; 2006

[10] Lamb H. Hydrodynamics. New York, USA: Dover Publications; 1932

[11] Chandrasekhar S. Hydrodynamic and Hydromagnetic Stability. New York, USA: Dover; 1961

[12] Kifondis K, Plewa T, Scheck L, Janka HT, Muller E. Non-spherical core collapse supernovae. Astronomy and Astrophysics. 2006;**453**(2):661-U17

[13] Malik HK, Tyagi J, Sharma D. Growth of Rayleigh instability in a Hall thruster channel having dust in exit region. AIP Advances. 2019;**9**:055220

[14] Selig F, Wermund EG. Families of salt domes in the gulf coastal province. Geophysics. 1966;**31**(4):726-740

[15] Ghasemizad A, Zarringhalam H, Gholamzadeh L. The investigation of Rayleigh-Taylor instability growth rate in inertial confinement fusion. Journal of Plasma Fusion Research. 2009;**8**:1234-1238

[16] Betti R, Umansky M, Lobatchev V, Goncharov VN, McCrory RL. Hotspot dynamics and deceleration-phase Rayleigh-Taylor instability of imploding inertial confinement fusion capsules. Physics of Plasmas. 2001;**8**(12):5257-5267

[17] Mccrory RL, Verdon CP, Betti R, Goncharov VN. Growth rates of the ablative Rayleigh-Taylor instability in inertial confinement fusion. Physics of Plasma. 1998;**5**(5):1446-1454

[18] Zhu G, Shi P, Yang Z, Zheng J, Luo M, Ying J, et al. A new method to suppress the Rayleigh-Taylor instability in a linear device. Physics of Plasmas;**26**: 2019, 042107

[19] Neill SM, Pesyridis A. Modeling of supersonic combustion systems for sustained hypersonic flight. Energies. 2017;**10**:900

[20] Rani K, Sharma SC. Theoretical modelling of Kelvin Helmholtz

instability driven by an ion beam in a negative ion plasma. Progress in Electromagnetics Research B. 2016;**71**: 167-181

[21] Cohen IM, Kundu PK. Fluid Mechanics. 3rd ed. California, USA: Elsevier Academic Press; 2004

[22] Reynolds O. An experimental investigation of the circumstances which determine whether the motion of water shall be direct or sinuous and of the law of resistance in parallel channels. Proceedings of the Royal Society of London. 1883;**35**:84-99

[23] Lord Rayleigh FRS. On the stability, or instability, of certain fluid motions. Proceedings of the London Mathematical Society. 1879;**s1-11**(1): 57-72

[24] Fjørtoft R. Application of integral theorems in deriving criteria of stability for laminar flows and for the baroclinic vortex. Geofys Publ. Oslo. 1950;**17**(6): 1-52

[25] Drazin PG, Reid WH. Hydrodynamic Stability. Cambridge, UK: Cambridge University Press; 1981

Chapter 3

Fluid Structure Interaction Analysis of Wind Turbine Rotor Blades Considering Different Temperatures and Rotation Velocities

Mayra K. Zezatti Flores, Laura Castro Gómez and Gustavo Urquiza

Abstract

Wind energy is the clean energy source that has had the highest installation growth worldwide. This energy uses the kinetic energy in the airflow currents to transform it into electrical energy through wind turbines. In this chapter, a rotor of a 2 MW of power wind turbine installed in Mexico is analyzed considering the wind velocity data and temperatures at each season of the year on the zone for the analysis in Computational Fluid Dynamics (CFD); subsequently, a Fluid–Structure Interaction (FSI) analysis was carried out to know the stress of the blades. The results show a relationship between temperature, air density, and power.

Keywords: FSI, wind turbine, CFD

1. Introduction

1.1 Wind energy in Mexico

Mexico has great potential for generating electricity from renewable resources. At the end of 2018, 75.88% of the energy used in the Mexican Republic comes from fossil fuels: such as oil, coal, and natural gas. Renewable energy generation reached 17.29% (hydroelectric, biogas, photovoltaic, wind, geothermal, and bagasse) and 6.83% from other clean energies (nuclear, efficient cogeneration, and black liquor) as illustrated in **Figure 1**, where it is observed the percentage distribution of total generation. There are currently 45 wind farms located in the country's eastern region (Oaxaca), where 59% of the total installed capacity is concentrated [1]. Also, other regions such as the northeast, northwest, western, peninsular, and Baja California.

Wind energy is generated from the kinetic energy in air currents and is transformed into electrical energy through wind turbines. Some important aspects for its generation are wind velocity and direction (Coriolis force), height, and temperature [2]. A minimum wind velocity of 3–5 m/s is required to start the rotor's rotation until reaching its maximum power.

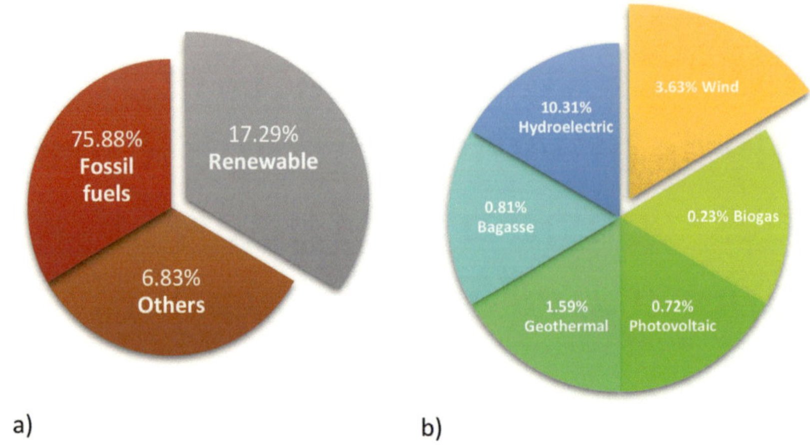

Figure 1.
(a) Distribution of total generation electricity in 2018 (b) renewable energy generation.

In the present work, a rotor of a wind turbine with a capacity of 2 MW is studied considering the real conditions of wind velocity at the temperatures presented in the year, first a Computational Fluid Dynamics (CFD) analysis, and subsequent a Fluid–Structure Interaction (FSI) analysis to know the stress of the blades of a turbine installed in a wind farm in the north of the country.

1.2 State of art

The rotor of a wind turbine is one of the most important components, for which it has been the object of study by some authors: Abolfazl Pourrajabian, carried out a study of the effect that air density has with respect to the torque of a small wind turbine of two meters in diameter, finding that the air density decreases as altitude increases, as well as the rotor torque [3]. Dong-Hoon Kim, studied a 5 MW turbine by means of fluid–structure, modeling in 2D only 1/3 of the rotor with different radial and longitudinal amplitudes of the rotary domain, considering a lineal composite material, obtaining the stress for a redesign of this [4]. Liping Dai, carried out a numerical study of the fluid structure of a Tjæreborg wind turbine, with a rotor diameter of 61.1 m, modeling a cylinder with the three blades to determine the behavior of deflection in a structural analysis, due to the effect of the velocity of wind with different angles of the YAW bearing [5]. Lanzafame analyzed a micro rotor of a horizontal axis turbine, to validate a BEM model in one dimension, comparing it with a CFD analysis performed in three dimensions; as a cylinder and two blades, and an experimental model, finding that the errors between simulated results in its power curve and the experimental data were less than 6% for all simulations [6]. With the previously analyzed studies, the characteristics, and conditions to carry out our case study were determined, such as domain dimensions, meshing with surface elements, turbulence model, wind velocity range, as well as the importance of determining the pressures. to perform a fluid–structure analysis.

2. Methodology

A wind turbine's operation is characterized by its power curve that indicates the range of wind velocities in which it can be operated and the one it generates.

(Eq. (1)) shows that the wind power depends on the swept area or the rotor exposed to a flow, as illustrated in **Figure 2**, on the fluid density and the wind

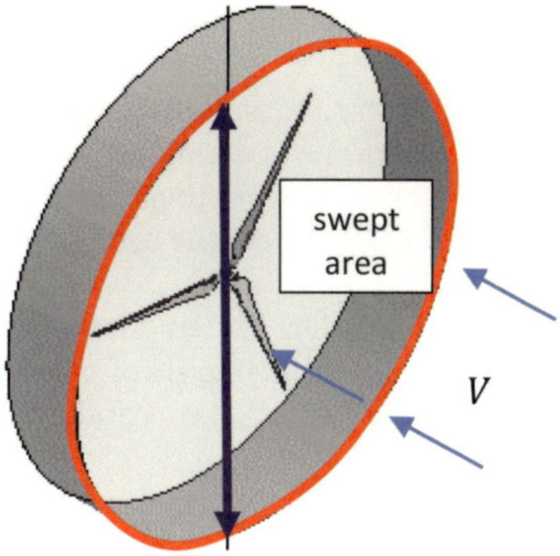

Figure 2.
Swept area of the blades of a wind turbine.

velocity. As Albert Betz demonstrated, wind energy is not used 100%; the limit he established was 59% for an ideal rotor. Additionally, the machine has mechanical, electrical, and aerodynamic losses [2, 7–9].

$$P = \frac{1}{2}\rho A V^3 \tag{1}$$

The Fluid–Structure Interaction (FSI) is the coupling of fluid and structural analysis. It considers the pressure or temperature of a CFD analysis and the direct consequences of this load on the structural analysis.

Two different software interact in the development of the fluid–structure analysis, where the independent fluid analysis is performed, and the required results are exported to the structural mesh for the solution. In this type of study, the CFD evaluation is solved in Fluent; then, the results are exported and imported to structural analysis, as illustrated in **Figure 3**.

2.1 CFD analysis

This analysis is carried out in three stages: pre-processing, processing, and post-processing. For the pre-processing development, the rotor geometry was generated

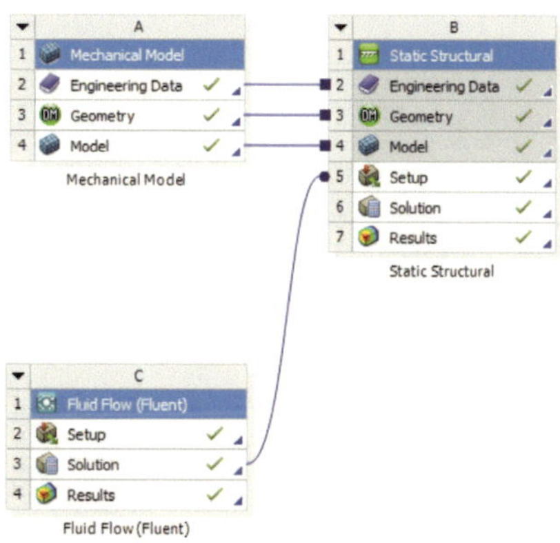

Figure 3.
Ansys diagram for the development of the FSI analysis.

in a CAD program from the information shown in **Table 1**, exported to another CAE program for the discretization.

Two domains were generated: rotary and stationary. **Figure 4** shows its dimensions, which were established according to the rotor diameter.

The discretization was carried out using an unstructured tetrahedral mesh. A refined mesh was realized in the zones near the rotatory interface and over the blade [10].

Were realized four meshes with different element sizes to perform an analysis of the independence of grid. This analysis consists of simulating at the same boundary conditions with a velocity inlet at 13 m/s and evaluating the torque results in these meshes.

It was calculated the percentage of relative error based on the torque result according to (Eq. (2)), considering the exact value of the one reported by the manufacturer in its power curve for a speed of 13 m/s, showing the values calculated in **Table 2**. According to the results, the mesh used has a relative error of 3.76%, illustrated in **Figure 5**, which shows the mesh in both domains.

$$Relative\ error = \frac{\left|V_{exact} - V_{approximate}\right|}{V_{exact}} \tag{2}$$

The definition of the physical models and solver configuration of the ANSYS Fluent software are shown in **Table 3**.

Statistical studies of the frequency for wind velocity and temperature were carried out (illustrated in **Figure 6**). The values of wind velocity are shown in the

Wind turbine	Characteristic
Diameter	87 m
Swept area	5,945 m^2
Rotation velocity	9–19 rpm
Profile	FFA + W3
Blade length	42.5 m
Design life	20 years

Table 1.
Rotor characteristics.

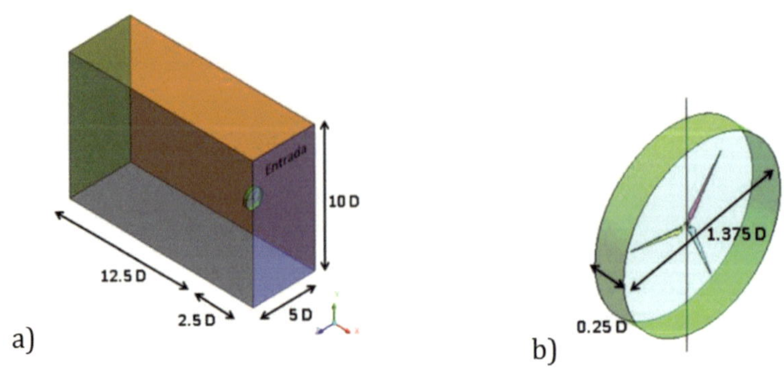

Figure 4.
Domain dimensions (a) stationary (b) rotary.

Type of mesh	Torque (N/m)	Relative error (%)
Reference (V_{exacto})	1,981,000	
Mesh1	1,647,193	16.85%
Mesh 2	1,696,181	14.38%
Mesh 3	1,906,495	3.76%
Mesh 4	1,865,284	5.84%

Table 2.
Calculation of the relative error.

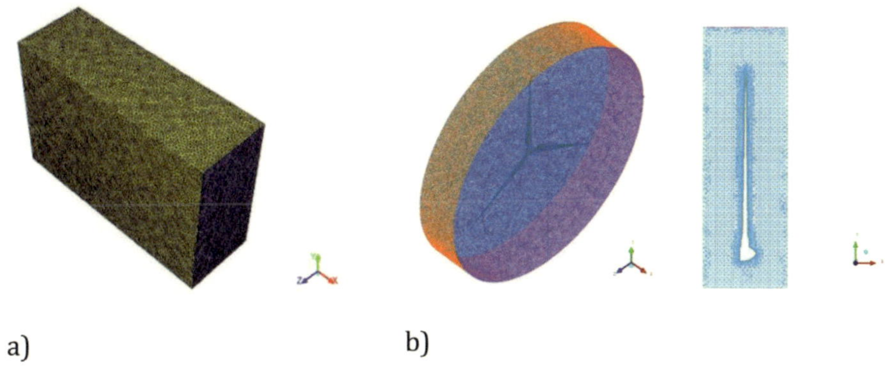

Figure 5.
Meshed domain (a) stationary (b) rotary.

Configuration	Characteristic
Type of analysis	Stationary
Turbulence model	kω-sst
Interfaces	input, output, outline
Solution method	SIMPLE
Boundary conditions	Vel inlet, pressure output, wall, symmetry

Table 3.
Solver configuration.

red box of the histogram of the frequency. On the other hand, for the temperature, the data were obtained by taking an average for the months of each season of the year: spring 20°, summer 38°, autumn 15° and winter −5 ° C, verifying them with the maps of the National Meteorological Service of the same year during the months of each season and considering these temperature values for the properties of air in the CFD analysis.

Ten analyzes were carried out for each temperature value in a wind velocity range of 4–13 m/s. As shown in **Table 4**, using the characteristics described above, obtaining the torque calculates the mechanical power with the (Eq. (3)) and its respective angular velocity.

$$P = M\,\omega \tag{3}$$

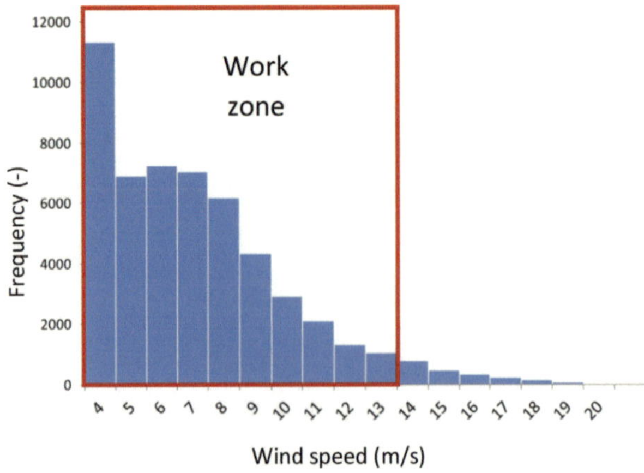

Figure 6.
Histogram of frequency of wind velocity during one year of operation.

Wind velocity (m/s)	Temperature (°C)
4	−5, 15, 20, 38
5	−5, 15, 20, 38
6	−5, 15, 20, 38
7	−5, 15, 20, 38
8	−5, 15, 20, 38
9	−5, 15, 20, 38
10	−5, 15, 20, 38
11	−5, 15, 20, 38
12	−5, 15, 20, 38
13	−5, 15, 20, 38

Table 4.
Operating conditions.

The post-processing of the results is presented in section 3.

2.2 FEM analysis

The rotor geometry is illustrated in **Figure 7a**, which was generated with surfaces following the data in **Table 1**, and **Figure 7b** demonstrates the thickness [11] imposed on the blades from root to the tip from 50 to 8 mm across the surface.

An analysis was performed with three different element sizes as seen in **Table 5**, with a rotational velocity of 19 rpm as a boundary condition to verify the maximum stress.

According to the results, it was decided to take mesh no.1 since the change in the data obtained is less than 5% with respect to mesh no.3. The numerical model of the rotor used is shown in **Figure 8**, which was carried out with second order elements shell281 [12].

The material of the hub and the blade was considered as a linear composite material [4]. As a boundary condition, support was imposed on the rotor hub and the rotational velocity in the direction of clockwise (Clockwise).

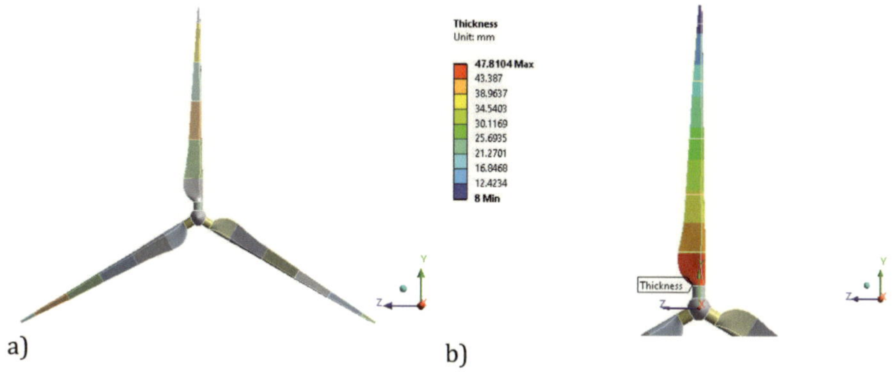

Figure 7.
Rotor characteristics (a) geometry (b) thickness distribution.

Mesh type	Number of nodes	Size element (mm)	Stress (MPa)
1	147,934	80	35.81
2	95,434	100	47.658
3	357,852	50	37.071

Table 5.
Von Mises stress results with different mesh sizes.

Figure 8.
Numerical rotor model.

Additionally, the evaluation's pressure results in CFD are included, imported to the structural analysis of each of the data obtained through interpolation of the ICEM mesh to the mesh generated in Ansys Workbench for the structural part.

The stress and strain results are shown in the next section.

3. Results

The CFD and FEM analysis simulation results are presented below, showing the mechanical power, pressures, and stress results, respectively.

3.1 CFD

Figure 9 shows four power curves, one curve for each temperature, and for each temperature ten analysis was performed for each value of wind speed in the range

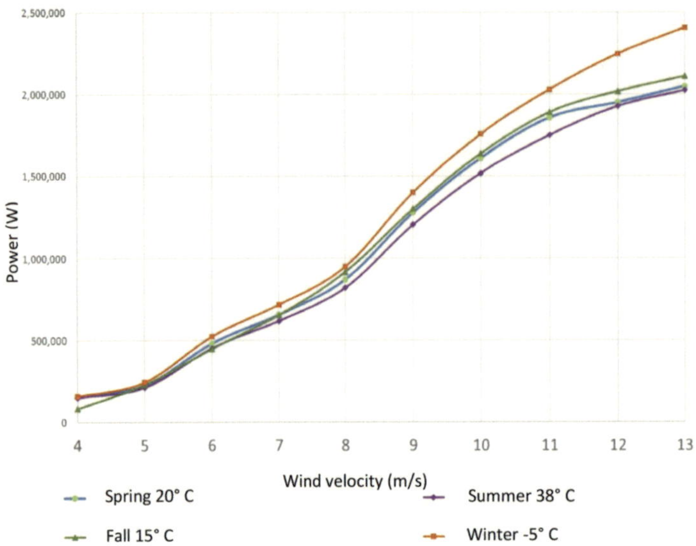

Figure 9.
Power at different temperatures.

of 4-13 m/s, obtaining a total of 40 simulations. The temperature directly affects the calculated power as seen in the figure, hence the air density [3].

When the temperature is increased, the air density decreases, and the power (Eq. (1)) shows that they are directly proportional.

The maximum power in the wind turbine area will be found during the winter when their temperature is −5 °C and the lowest during the summer that reaches 38 °C in the installed area.

The absolute pressure contour was obtained at all nodes on the rotor surface (**Figure 10**). The maximum stress is found at the edge of the profile outlet because the wind velocity is normal to the blade's plane, and as the profile rotates, it is the edge with the most significant impact.

3.2 FEM

An analysis was carried out at the different rotation velocitys in the operating range of 9–19 r.p.m. reported by the manufacturer. It can be noticed that the

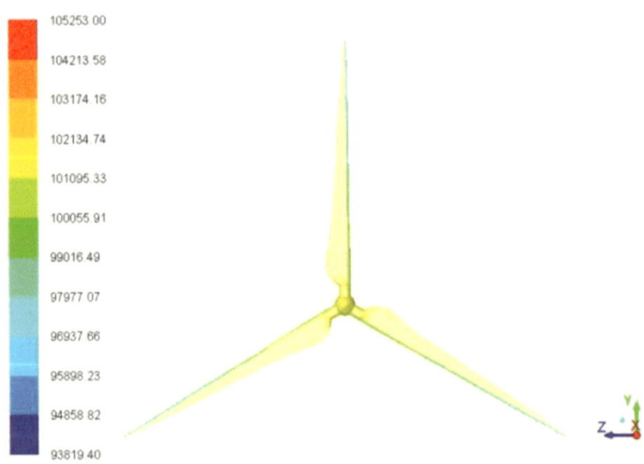

Figure 10.
Absolute pressure contour al blades (Pa).

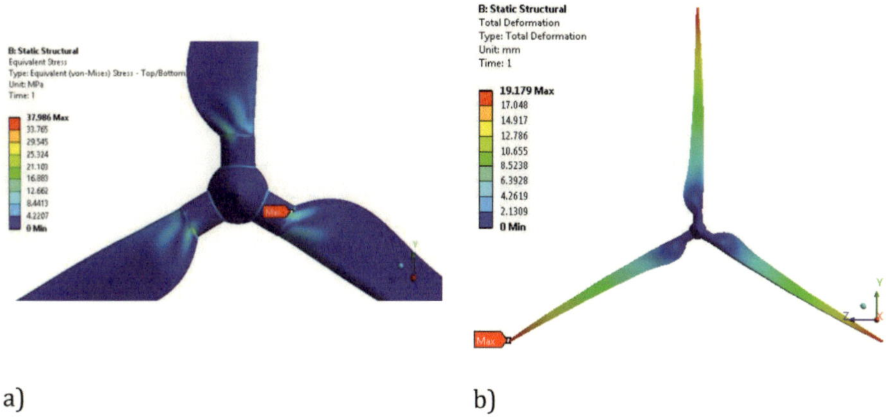

Figure 11.
Results at 19 r.p.m. (a) Von Mises stress state (b) maximum displacement.

contribution is minimal since the maximum stress of 8.5 MPa was obtained with the lower limit and 37.9 MPa with the upper limit.

The results at 19 r.p.m. are shown in **Figure 11**, in (a) the maximum stress distribution at the root of the blade in conjunction with the cylinder due to the change in thickness and b) the maximum displacement found at the tip of the blade.

3.3 FSI

The obtained pressures were exported to ANSYS Mechanical at each wind velocity for its structural analysis, adding the mentioned boundary conditions to obtain a stress state.

The state of stress at different temperatures in the predominant velocity range of 4-13 m/s show in **Figure 12**. It is observed that at 38 °C during the summer, the stresses are less than during the winter, at −5 °C. The centrifugal force contributes approximately 30% of the total effort with the pressure exerted by the air on the rotor's surface.

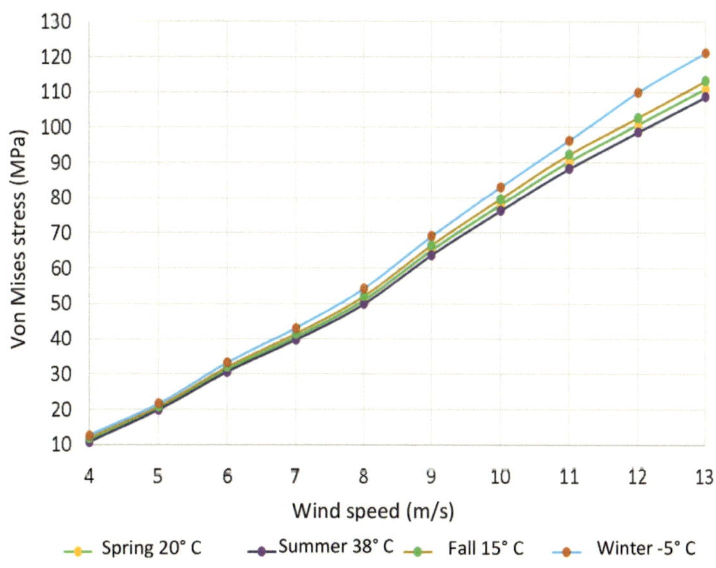

Figure 12.
Von Mises stresses at different temperatures.

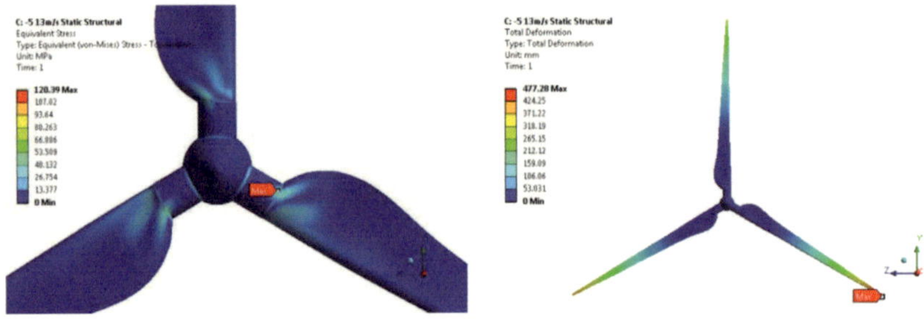

a) b)

Figure 13.
Rotor results at 13 m/s. (a) Von Mises stress state (b) maximum displacement.

Figure 13(a) shows the Von Mises stress distribution at a maximum wind velocity of 13 m/s, in which the blade root zone is affected by the bending of the blade tip, and on (b) illustrates the maximum displacement located at the tip of the blade, where the effect of the rotation and the pressure of the air entering to the surface affects the area with the minimum thickness.

4. Conclusions

The CFD methodology used in the present work was validated by comparing the results between the calculated power with the manufacturer power, obtaining a valid approximation to the real phenomenon, taking into account the ideality of the case under study, in addition to not considering the aerodynamic, mechanical and electrical losses of the wind-power generator.

Was performed ten CFD simulations for each value of temperature to obtain the pressures for the structural analyses. It was demonstrated that the mechanical power has a directly proportional relationship with the temperature and with the air density, with which it is concluded that the maximum power in the year is generated during the winter, average power in summer and autumn, and a minimum power in the summer in the installation area.

The fluid–structure analyses include the forty simulations performed for CFD's and the stress for the structural analysis at different rotation velocities. The maximum stress was found at 13 m/s, lower for 38 °C and higher for −5 °C, located mainly at the blade's root due to the change in thickness and rotation velocity and the pressure exerted by the air. Which 120 MPa does not exceed the yield stress of 240 MPa.

Author details

Mayra K. Zezatti Flores, Laura Castro Gómez* and Gustavo Urquiza
Autonomous University of Morelos State (UAEM), Morelos, Mexico

*Address all correspondence to: lauracg@uaem.mx

References

[1] SENER, "Reporte de Avance de Energías Limpias Primer Semestre 2018," México, 2018.

[2] W. Tong et. al., "Wind Power Generation and Wind Turbine Design," *WIT press,* pp. ISBN 978-1-84564-205-1, 2010.

[3] Abolfazl Pourrajabian et. al., "Effect of air density on the performance of a small wind turbine blade: A case study in Iran," *Wind Eng. Ind. Aerodyn.,* pp. 1-10 p. DOI: doi.org/10.1016/j.jweia.2014.01.001, 126, 2014.

[4] Dong-Hoon Kim et. al., "Optimization of 5-MW wind turbine blade using fluid structure interaction analysis," *Mechanical Science and Technology,* pp. 725-732 p. DOI: doi.org/10.1007/s12206-017-0124-2, 2017.

[5] Liping Dai, et. al., "Analysis of wind turbine blades aeroelastic performance under yaw conditions," *Journal of Wind Engineering & Industrial Aerodynamics,* vol. 171, pp. 237-287 DOI: doi.org/10.1016/j.jweia.2017.09.011, 2017.

[6] Lanzafame et. al., "Wind turbine CFD modeling using a correlation-based transitional model," *Renewable Energy,* vol. 52, pp. 31-39 DOI:doi.org/10.1016/j.renene.2012.10.007, 2013.

[7] M. F. Voneschen, Introducción a la teoría de las turbinas eólicas, España: LA VERITAT ISBN 3-7625-2700-8 , 2009.

[8] A. Betz, La energía eólica y su aprovechamiento mediante molinos de viento, traducido por Manuel Franquesa Voneschen, 2012.

[9] A. P. Schaffarczyk, "Introduction to wind turbine aerodynamics", Alemania: Springer DOI 10.1007/978-3-642-36409-9, 2014.

[10] J.M. O'Brien et. al., "Horizontal axis wind turbine research: A review of commercial CFD, FE codes and experimental practices," *Progress in Aerospace Sciences,* pp. 1-24 p. DOI: doi.org/10.1016/j.paerosci.2017.05.001, 2017.

[11] Jin Chen, et. al., "Structural optimization study of composite wind turbine blade," *Materials and Design,* pp. 247-255 p. DOI: doi.org/10.1016/j.matdes.2012.10.036, 2013.

[12] *Ansys user's guide,* 2019.

Chapter 4

Dynamics of Rayleigh-Taylor Instability in Plasma Fluids

Sukhmander Singh

Abstract

The chapter discusses the evolution of Rayleigh-Taylor instability (RTI) in ordinary fluids and in a plasma fluid. RT instability exits in many situations from overturn of the outer portion of the collapsed core of a massive star to laser implosion of deuterium-tritium fusion targets. In the mixture of fluids, the instability is triggered by the gravitational force acting on an inverted density gradient. The motivation behind the study of the instability has been explored by discussing the applications of RT instability. The basic magnetohydrodynamics equations are used to derive the dispersion relation (for an ordinary fluid and plasmas) for two fluids of unequal densities. The conditions of the growth rate of the instability and the propagating modes are obtained by linearizing the fluid equations. The perturbed potential is found to increase with the plasma parameters in a Hall thruster.

Keywords: instabilities, plasma, Navier-Stokes, growth rate, Hall thruster

1. Introduction

Flow instabilities are used to increase the heat and mass transfer rates as well as to fuse the fluids of dissimilar properties (viscosity, elasticity, density, etc.). In other technological applications, these instabilities are accountable to unstable the multilayer and free-surface flows. Multilayer flows are used in coating processes and lubricated pipeline transport. The presence of the instabilities in the system leads to nonuniform film thickness and defects, where good optical finishing and smooth edges are required by the industry, which further leads to poor product quality. Suppression of these instabilities has been a major task from a long time by the researchers to improve the product quality [1, 2]. Rayleigh-Taylor (RT) instability takes place when a lighter fluid supports a heavy fluid, then any perturbation of the interface grows and leads to spikes of the heavier fluid penetrating into the lighter one and the interface becomes unstable. The contact discontinuity between the two fluids is unstable to perturbations that grow by converting potential energy to kinetic energy, causing bubbles of the low-density fluid to rise, and spikes of the high-density fluid to sink. If the light fluid is above the heavy fluid, the interface is stable. In a magnetized plasma, the Rayleigh-Taylor instability can occur because the magnetic field acts as a light fluid supporting a heavy fluid (the plasma).

In curved magnetic fields, the centrifugal force on the plasma due to the charged particle motion along the curved field lines acts as an equivalent gravity force. When forces associated with the density gradient and gravity oppose each other, the RT instability sets in [3, 4]. The box of fluid shown in **Figure 1** is now filled with

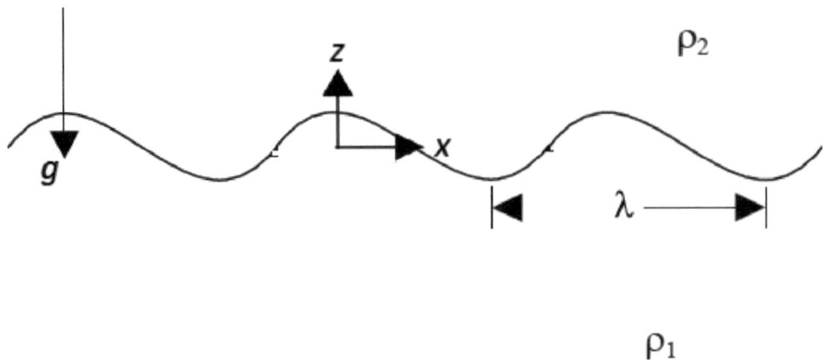

Figure 1.
Two fluids inside of a large box.

two incompressible fluids of differing densities, separated by an interface with a perturbation imposed as shown in **Figure 1**. Here, RTI is seen to play a wider role in many branches of science from astrophysical systems to industries.

2. Review of status of research

This instability occurs in many interesting physical situations, such as implosion of inertial confinement fusion capsules, core collapse of supernovae, or electromagnetic implosions of metal liners. The Rayleigh-Taylor problem was first studied by Lord Rayleigh in 1883 and Sir G.I. Taylor in 1950 [3]. Taylor used the theory of linearization for the small oscillations at the interface and obtained an exponential growth rate. Chandrasekhar, in 1961, studied the magnetic field case analytically for the fluids that are incompressible, inviscid, and have zero resistivity. Qin et al. [5] reported the synthesis of chains of metal nanoparticles with well-controlled particle sizes and spacing induced by the Rayleigh instability. Bychkov et al. [6] derived the dispersion relation for the internal waves and the RT instability in a nonuniform unmagnetized quantum plasma with a constant gravitational field. They have shown that the quantum effects always play a stabilizing role for the RT wave instability. Cao et al. [7] studied the RT instability incorporating the quantum magnetohydrodynamic equations and solved the second-order differential equation under different boundary conditions with quantum effects. Khomenko et al. [8] modeled the growth rate of the instability and the evolution of velocity and magnetic field vector in the prominence plasma (closer to Sun's surface) under the presence of neutral atoms. Diaz et al. [9] derived the criterion for the growth rate of the RT instability in partially ionized plasma using single fluid theory. Ibrahim and Marshall theoretically investigated the impact of velocity profile on RTI within the jet to examine the effects of its relaxation on intact length [10]. Carlyle and Hillier experimentally verified that stronger magnetic fields can suppress the growth of the rising bubbles of the RTI [11]. Litvak and Fisch derived the necessary instability conditions of azimuthally propagating perturbations in a Hall thruster plasma [12]. Recently, investigators derived the dispersion for the Rayleigh-Taylor instabilities in a Hall thruster using the two - fluid theory [13, 14]. Shorbagy and Shukla investigated the RT instability in a nonuniform multi-ion plasma in a Hall thruster to obtain the growth rate of the instability [15]. Ali et al. [16] derived the modified dispersion relation for the Rayleigh-Taylor instability under the quantum corrections incorporating the terms of Fermi pressure and the Bohm potential force.

3. Basic fluid equations and Bernoulli's theorem

First, we consider the two simple fluids separated by a smooth interface to derive the dispersion relation. Let us assume that in each separate region, the density is constant. The coordinate x is in the horizontal, z in the vertical, and y is going into the page. We consider a flow in the x-direction, which in the lower half-space ($z < 0$) has density ρ_1, whereas in the upper half-space ($z > 0$) has density ρ_2. In addition there can be a homogeneous gravitational field g pointing into the negative z-direction. We write the basic fluid equations for the ion and electron fluids as Navier-Stokes equations for an incompressible fluid are

$$\frac{\partial \rho}{\partial t} + \vec{\nabla} \cdot \left(\rho \vec{v}\right) = 0 \tag{1}$$

$$\frac{d\vec{v}}{dt} = \frac{\partial \vec{v}}{\partial t} + \left(\vec{v} \cdot \vec{\nabla}\right)\vec{v} = -\frac{\vec{\nabla} P}{\rho} + g + \eta \nabla^2 \vec{v} \tag{2}$$

Here, we have used total time derivative. Partial time derivative keeps an eye on a point and represents the rate of velocity change at that point. Total time derivative keeps an eye on fluid element and measures its velocities at t and $t + \Delta t$.

Let us consider the fluid is inviscid, so that we take viscosity $\eta = 0$. We also assume that the fluid is irrotational, that is $\vec{\nabla} \times \vec{v} = 0$. Then the term $\left(\vec{v} \cdot \vec{\nabla}\right)\vec{v}$ reduces to $\frac{1}{2}\vec{\nabla} v^2$. The Stokes' theorem permits us to express the velocity in terms of gradient of scalar function, that is $\vec{v} = -\vec{\nabla}\phi$. The variable ϕ is called the scalar velocity potential of fluid. We rewrite gravity acceleration into a gradient of gravity potential $g = -\vec{\nabla}(gz)$. Eq. (2) can be rewritten in terms of scalar function ϕ under the above assumptions:

$$\vec{\nabla}\frac{\partial \phi}{\partial t} + \frac{1}{2}\vec{\nabla} v^2 = -\frac{\vec{\nabla} P}{\rho} - \vec{\nabla}(gz) \tag{3}$$

If the density remains constant in one region, we can write Eq. (3) as

$$\vec{\nabla}\left[\frac{\partial \phi}{\partial t} + \frac{1}{2}\left(\vec{\nabla}\phi\right)^2 + gz\right] = -\frac{\vec{\nabla} P}{\rho} \tag{4}$$

Now integrating the above equation in horizontal and vertical directions, we get unsteady equation for the Bernoulli theorem.

$$\frac{\partial \phi}{\partial t} + \frac{1}{2}\left(\vec{\nabla}\phi\right)^2 + gz + \frac{P}{\rho} = Const \tag{5}$$

That is, the total mechanical energy of the moving fluid comprising the gravitational potential energy of elevation, the energy associated with the fluid pressure, and the kinetic energy of the fluid motion remains constant.

Let $\vec{v}_1$ and $\vec{v}_2$ be the velocities of the fluid in the lower half-space ($z < 0$) and upper half-space ($z > 0$) respectively. Now, it is convenient to write velocities of fluid in terms of scalar velocity potential ϕ in both regions such that

$$\vec{v}_1 = -\vec{\nabla}\phi_1 \tag{6}$$

$$\vec{v}_2 = -\vec{\nabla}\phi_2 \tag{7}$$

For the incompressible fluid, Eq. (1) yields that $\vec{\nabla} \cdot \vec{v} = 0$. This also states that both fluids will satisfy the Laplace equation in both regions

$$\vec{\nabla}^2 \phi = 0 \tag{8}$$

The Bernoulli theorem state that quintiles $\rho \frac{\partial \phi}{\partial t} + \frac{1}{2}\rho\left(\vec{\nabla}\phi\right)^2 + \rho g z + P$ should be constant across the fluid, so that we have

$$\rho_1 \frac{\partial \phi_1}{\partial t} + \frac{1}{2}\rho_1\left(\vec{\nabla}\phi_1\right)^2 + \rho_1 g z_0 + P_1\Big|_{z=z_0} = \rho_2 \frac{\partial \phi_2}{\partial t} + \frac{1}{2}\rho_2\left(\vec{\nabla}\phi_2\right)^2 + \rho_2 g z_0 + P_2\Big|_{z=z_0} \tag{9}$$

To understand the interface, we must impose boundary conditions. First of all the vertical velocities of the fluids must match with the interface, so we impose the kinematic boundary condition. Now we need to introduce the location of the interface by assigning variable $z = z_0(t)$. Then $\frac{dz_0}{dt}$ will represent the velocity of the interface in the z-direction. In addition, at the interface, the velocity of both fluids must be continuous.

$$\frac{dz_0}{dt} = \left(\frac{\partial}{\partial t} + \left(\vec{v} \cdot \vec{\nabla}\right)\right)z_0 = \frac{\partial \phi_1}{\partial z}\Big|_{z=z_0} = \frac{\partial \phi_2}{\partial z}\Big|_{z=z_0} \tag{10}$$

Let us say that the pressure is continuous along the interface, that is $P_1 = P_2$. Then Eq. (9) leads to

$$\rho_1 \frac{\partial \phi_1}{\partial t} + \frac{1}{2}\rho_1\left(\vec{\nabla}\phi_1\right)^2 + \rho_1 g z_0 = \rho_2 \frac{\partial \phi_2}{\partial t} + \frac{1}{2}\rho_2\left(\vec{\nabla}\phi_2\right)^2 + \rho_2 g z_0 \tag{11}$$

4. Asymptotic boundary conditions at far field

We are looking for changes only on the interface at $z = z_0$, therefore the velocity potentials and their derivatives must vanish at the boundaries, that is $\phi_1 \to 0$ as $z \to -\infty$ and $\phi_2 \to 0$ as $z \to \infty$.

5. Linear analysis

Eq. (9) contains a nonlinear term $\rho_1\left(\vec{\nabla}\phi_1\right)^2$ of the second order. If the amplitude is chosen to be much smaller than the wavelength of the instability, the equations of motion can be linearized. We assume that all the perturbed quantities and various derivatives such as ϕ and $\vec{\nabla}\phi$ are very small. In other words $\frac{\partial \phi_1}{\partial z}\Big|_{z=z_0} \sim \frac{\partial \phi_1}{\partial z}\Big|_{z=0}$. Hence, the difference in second order derivatives will be much smaller. Now we impose all boundary conditions at $z = 0$, which yields the following set of equations,

$$\vec{\nabla}^2 \phi_1 = 0 \tag{12}$$

$$\vec{\nabla}^2 \phi_2 = 0 \tag{13}$$

$$\frac{dz_0}{dt} = \left.\frac{\partial \phi_1}{\partial z}\right|_{z=0} = \left.\frac{\partial \phi_2}{\partial z}\right|_{z=0} \tag{14}$$

$$\rho_1 \frac{\partial \phi_1}{\partial t} + \rho_1 g z_0 \bigg|_{z=0} = \rho_2 \frac{\partial \phi_2}{\partial t} + \rho_2 g z_0 \bigg|_{z=0} \tag{15}$$

6. Eigenvalue solution

Let us consider all the perturbed variables ϕ and z_0 to have oscillating behavior such that $\phi = \phi_0 \exp\left[i(kx - \omega t)\right]$ and should satisfy the Laplace Eq. (12). This implies

$$\frac{\partial^2 \phi_1}{\partial z^2} = k^2 \phi_1 \tag{16}$$

The general solution of Eq. (16) is written as

$$\phi_1(z) = A \exp(kz) + B \exp(-kz), \tag{17}$$

The above two solutions must satisfy the boundary conditions such that $\phi_1 \to 0$ as $z \to -\infty$ and $\phi_2 \to 0$ as $z \to \infty$. So, we need to discard the unsatisfactory part of the solutions of Eq. (16) taking into account the boundary conditions. Therefore, z dependence goes as $\phi_1(z) \propto A \exp(kz)$ and $\phi_2(z) \propto B \exp(-kz)$. We note that the eigenfunction decreases exponentially on either side of the interface and the perturbation of wave number k penetrates to a depth of order $\frac{1}{k} = \frac{\lambda}{2\pi}$.

This solution further leads to the following form in Fourier mode,

$$\phi_1(x, z, t) = \phi_{01} \exp(kz) \exp\left[i(kx - \omega t)\right] \tag{18}$$

$$\phi_2(x, z, t) = \phi_{20} \exp(-kz) \exp\left[i(kx - \omega t)\right] \tag{19}$$

$$z_0(x, t) = z_{00} \exp\left[i(kx - \omega t)\right] \tag{20}$$

Here ϕ_{01}, ϕ_{02}, and z_{00} are the amplitude of the modes. By substituting these solutions into Eq. (14), we obtain the boundary conditions at the interface.

$$k\phi_{01} = -k\phi_{02} = -i\omega z_{00} \text{ at } z = 0 \tag{21}$$

Then Eq. (20) changes into the form

$$z_0(x, t) = \frac{ik\phi_{01}}{\omega} \exp\left[i(kx - \omega t)\right] \tag{22}$$

Eq. (15) gives

$$-i\omega\rho_1\phi_{10} + \rho_1 g \frac{ik\phi_{01}}{\omega} = -i\omega\rho_2\phi_{20} + \rho_2 g \frac{ik\phi_{01}}{\omega} \tag{23}$$

Using Eq. (21) in Eq. (23) results in

$$-i\omega\rho_1\phi_{10} + \rho_1 g \frac{ik\phi_{01}}{\omega} = i\omega\rho_2\phi_{10} + \rho_2 g \frac{ik\phi_{01}}{\omega} \tag{24}$$

Since the perturbed quantity $\phi_{01} \neq 0$, the possible nontrivial solution of Eq. (25) gives the dispersion relation for small perturbations of the wave as below,

$$\omega^2 = \frac{g(\rho_1 - \rho_2)k}{\rho_1 + \rho_2} = A_t gk \tag{25}$$

Eq. (25) contains complete information about the linear stability of the two superposed fluid layers of different densities. The Atwood number $A_t = \frac{(\rho_1 - \rho_2)}{\rho_1 + \rho_2}$ is a dimensionless number in fluid dynamics used to study the hydrodynamic instabilities in unequal density flows. Since the dispersion relation Eq. (25) is quadratic in ω, it has two real or complex conjugate roots depending on the values of the densities of the fluids. Here, we will discussion different cases.

6.1 First case: capillary-gravity waves ($\rho_2 = 0$)

$$\text{Hence } \omega = \sqrt{gk} \text{ and } V_{ph} = \sqrt{\frac{g}{k}} \tag{26}$$

It is classical dispersion relation for gravity-capillary waves in deep water [17, 18]. These are also called short gravity waves. In this category the longer waves travel faster. Any initial disturbance may be regarded as the superposition of waves of a broad spectrum of lengths. The above relation then says that waves of different lengths will eventually separate, that is, disperse. This phenomenon is called dispersion, hence above relations are also known as the dispersion relation.

6.2 Second case: propagating modes ($\rho_1 > \rho_2$)

If the lighter fluid is supported by heavier fluid, that is, $\rho_1 > \rho_2$, then solutions of the equation leads to two waves with constant amplitude propagating in opposite directions with phase velocity $\frac{\omega_\pm}{k}$ with $\omega_\pm = \sqrt{\frac{g(\rho_1 - \rho_2)k}{\rho_1 + \rho_2}}$. Then the interface is stable and will only oscillate when perturbed. The phase velocity is given by

$$V_{ph} = \frac{\omega_\pm}{k} = \sqrt{\frac{g\left(\frac{\rho_1}{\rho_2} - 1\right)}{k\left(\frac{\rho_1}{\rho_2} + 1\right)}} \tag{27}$$

Figure 2 shows the variations of phase velocity of RT instability with (a) density ratio and (b) wave number respectively.

6.3 Third case: Rayleigh-Taylor instability ($\rho_2 > \rho_1$)

The frequency of oscillations will be negative imaginary and unstable if $\rho_2 > \rho_1$, that is, when heavier fluid is supported by lighter fluid. Writing $\omega = i\gamma$ where γ is real and positive gives

$$\gamma = \pm\sqrt{\frac{g(\rho_2 - \rho_1)k}{\rho_1 + \rho_2}} \tag{28}$$

$$\gamma = \pm\sqrt{\frac{g\left(\frac{\rho_2}{\rho_1} - 1\right)k}{1 + \frac{\rho_2}{\rho_1}}} \tag{29}$$

Substituting the value of $\omega = i\gamma$ into Eq. (18), the amplitude grows exponentially with the perturbation and is given by

$$\phi_1(x, z, t) = \phi_{01} \exp(kz) \exp(ikx) \exp(\pm\gamma t) \tag{30}$$

The term $\exp(\gamma t)$ increases the amplitude of the oscillation exponentially as time progress. **Figure 3(a)** and **(b)** shows the variations of growth rate of RT instability with (a) density ratio and (b) wave number. The inverse of the growth rate $\gamma^{-1} = t_{char}$ is called the linear characteristic timescale of the RTI. In other words, characteristic timescale has to be the order of the lifetime of the plasma oscillations to observe the RT instability. On the other hand, if the linear characteristic timescale is much larger than the oscillation lifetime, the plasma instability would not be observed.

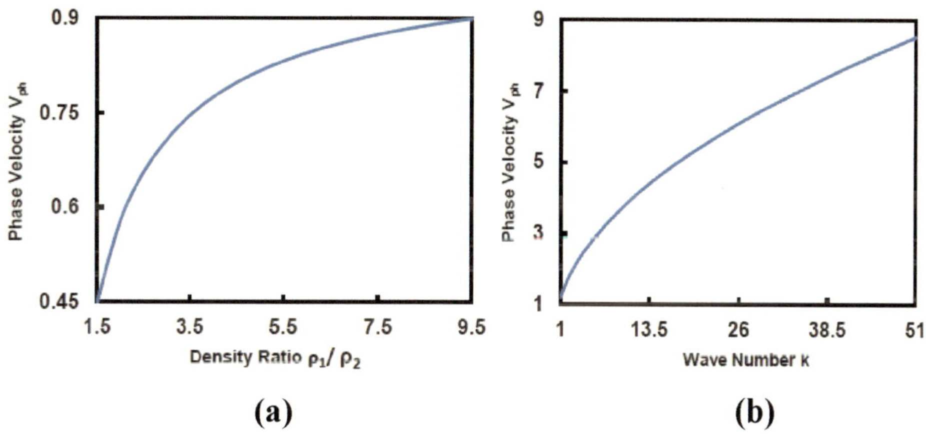

Figure 2.
Variation of phase velocity of RT instability with (a) density ratio and (b) wave number respectively.

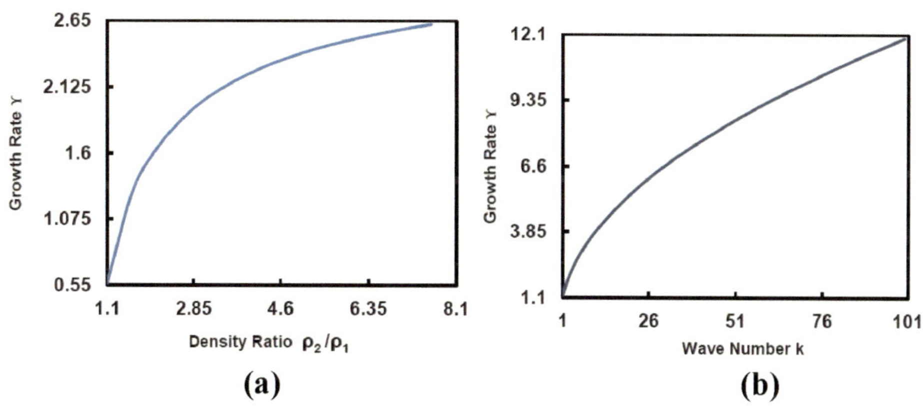

Figure 3.
Variation of growth rate of RT instability with (a) density ratio and (b) wave number respectively.

7. Rayleigh-Taylor instability in plasma thruster

In the previous section, the general idea of RT instability has been explored. Here we have derived the RT equation for a plasma fluid using two fluid theory. In a Hall thruster, the propellant (plasma) is ionized and then accelerated by electrostatic forces. It has high thrust resolution, so it is best suited for the adjustment of the location of the satellite onboard [19–27]. Let us consider a plasma with nonuniform density confined under the crossed electric and magnetic fields.

Figure 4 shows the typical diagram of a Hall plasma thruster [26]. RT instability is common in Hall thrusters. Studies show that Rayleigh instability is driven by the presence of gradients in axial density, magnetic field, and velocity of the plasma species. Here we deduce a Rayleigh equation under the presence of ion temperature and check the variations of perturbed potential with plasma parameters.

7.1 Theoretical model for RTI in plasma

We consider plasma comprising of ions and electrons immersed in a magnetic field $\vec{B} = B\hat{z}$. The magnetic field is strong enough so that only electrons get magnetized, but the ions remain unaffected due to their Larmor radius being much larger than the dimension of the thruster. These trapped electrons (due to crossed fields) drift in azimuthal direction along the annular channel [24]. The applied electric field $\vec{E}$ is along the x-axis (axis of the thruster) and the magnetic field $\vec{B}$ is taken along the z-axis (along the radius of the thruster). Hence, the azimuthal dimension is along the y-axis. We use $\Omega_z = \frac{eB}{m}$ as the electron gyro frequency and $u_0 = -\frac{E_0}{B}\hat{y}$ as the initial drift of the electrons [14–17] and write the continuity equation and equation of motion for plasma species as

$$\frac{\partial n_i}{\partial t} + \vec{\nabla} \cdot \left(\vec{v}_i n_i\right) = 0 \tag{31}$$

$$\left(\frac{\partial}{\partial t} + \vec{v}_i \cdot \vec{\nabla}\right)\vec{v}_i = \frac{e\vec{E}}{M} - \frac{\vec{\nabla} p_i}{M n_i} \tag{32}$$

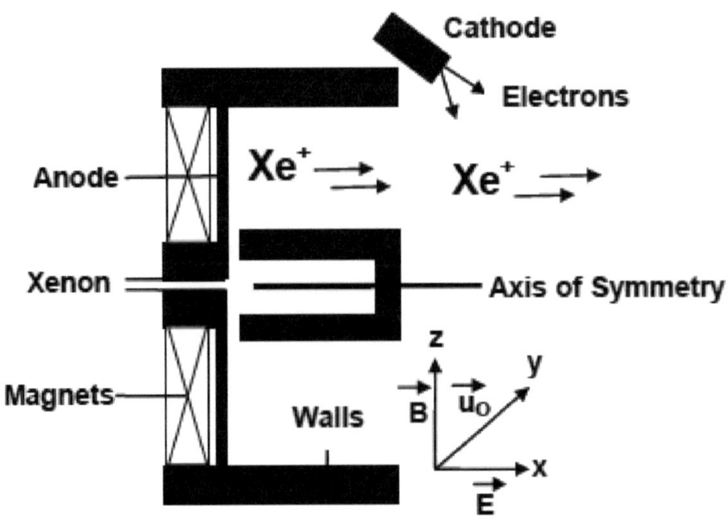

Figure 4.
Typical diagram of a Hall plasma thruster.

$$\frac{\partial n_e}{\partial t} + \vec{\nabla} \cdot \left(\vec{v}_e n_e \right) = 0 \tag{33}$$

$$\left(\frac{\partial}{\partial t} + \vec{v}_e \cdot \vec{\nabla} \right) \vec{v}_e = \frac{\vec{\nabla}\phi}{m} - \left(\vec{v}_e \times \vec{\Omega}_z \right) \tag{34}$$

We use the linearized form of the above equations for small perturbations of the ion and electron densities, their velocities, and electric field. We write perturbed densities of ions (electrons) by n_{i1} (n_{e1}) velocities by $\vec{v}_{i1}$ ($\vec{v}_{e1}$). The unperturbed electrons' drift is u_0 in the y-direction. The unperturbed density (electric field) is taken as $n_0(E_0)$ and the perturbed value of the electric field is taken as $\vec{E}$ (corresponding potential ϕ). Hence, the linearized form of Eqs. (31)–(34) reads

$$\frac{\partial n_{i1}}{\partial t} + v_{ix1}\frac{\partial n_0}{\partial x} + n_0\left(\vec{\nabla} \cdot \vec{v}_{i1} \right) = 0 \tag{35}$$

$$\frac{\partial \vec{v}_{i1}}{\partial t} = \frac{e}{M}\vec{E} - \frac{\vec{\nabla} p_{i1}}{n_0 M} \tag{36}$$

$$\frac{\partial n_{e1}}{\partial t} + u_0\frac{\partial n_{e1}}{\partial y} + n_0\left(\vec{\nabla} \cdot \vec{v}_{e1} \right) + v_{ex1}\frac{\partial n_0}{\partial x} = 0 \tag{37}$$

$$\frac{\partial \vec{v}_{e1}}{\partial t} + u_0\frac{\partial \vec{v}_{e1}}{\partial y} = \vec{\nabla}\phi - \left(\vec{v}_{e1} \times \vec{\Omega}_z \right) \tag{38}$$

The unperturbed ions' velocity v_0 is taken zero here for the case of simplification. We are looking for oscillating solution of the above equations that should vary as $f = f_0 \exp\left(i\omega t - ik_y y\right)$. The ion thermal velocity can be written as $V^2{}_{thi} = \frac{Y_i T_i}{M}$. With the help of Eqs. (35) and (36) we obtain the following expression for the perturbed ion density in terms of the perturbed electric potential ϕ:

$$n_{i1} = \frac{e n_0}{M\left(\omega^2 - V^2{}_{thi} k_y{}^2\right)} \left(k_y{}^2 \phi - \frac{\partial^2 \phi}{\partial x^2} - \frac{Y_i T_i}{e n_0} \frac{\partial^2 n_0}{\partial x^2} \right) \tag{39}$$

Eq. (38) provides the velocity components of electron

$$i\left(\omega - k_y u_0\right) v_{ex1} = \frac{e}{m}\frac{\partial \phi}{\partial x} - \Omega_z v_{ey1} \tag{40}$$

$$i\left(\omega - k_y u_0\right) v_{ey1} + v_{ex1}\frac{\partial u_0}{\partial x} = -ik_y\frac{e}{m}\phi + \Omega_z v_{ex1} \tag{41}$$

In the above equations, the coordinate x lies in the interval $0 < x < d$, where d is the channel length. Let us define $\omega - k_y u_0$ by $\hat{\omega}$ in the above set of expressions. Further we readily obtain from the above equations

$$v_{ex1} = \frac{\frac{e}{m} i\hat{\omega}\frac{\partial \phi}{\partial x} + \frac{e}{m} ik_y \Omega_z \phi}{\Omega_z{}^2 - \hat{\omega}^2 - \Omega_z \frac{\partial u_0}{\partial x}} \tag{42}$$

$$v_{ey1} = \frac{e}{m\Omega_z}\frac{\partial \phi}{\partial x} + \frac{\frac{e}{m}\hat{\omega}^2 \frac{\partial \phi}{\partial x} + \frac{e}{m}\hat{\omega} k_y \Omega_z \phi}{\Omega_z\left(\Omega_z{}^2 - \hat{\omega}^2 - \Omega_z \frac{\partial u_0}{\partial x}\right)} \tag{43}$$

The electron cyclotron frequency is almost $\Omega_z \sim 10^8$/s (corresponding to 200 Gauss magnetic field). Generally, Ω_z is much larger than the frequency of the oscillations. Therefore, under the condition $\Omega_z >> \omega, k_y u_0, \frac{\partial u_0}{\partial x}$, the velocity components of electrons are reduced into the form

$$v_{x1} = \frac{ie}{m\Omega_z^2}\left[\hat{\omega}\frac{\partial\phi}{\partial x} + \Omega_z k_y \phi + \frac{\partial u_0}{\partial x} k_y \phi\right] \tag{44}$$

$$v_{y1} = \frac{e}{m\Omega_z}\left[\frac{\partial\phi}{\partial x} + \frac{k_y\hat{\omega}\phi}{\Omega_z} + \frac{k_y\hat{\omega}\phi}{\Omega_z^2}\frac{\partial u_0}{\partial x}\right] \tag{45}$$

The electron continuity equation gives the perturbed electron density n_e with the help of Eqs. (44) and (45)

$$n_{e1} = \frac{en_0}{m\Omega_z^2}\left[\left(k_y^2\phi - \frac{\partial^2\phi}{\partial x^2}\right) + \frac{k_y}{\omega - k_y u_0}\left(\Omega_z\frac{\partial}{\partial x}\ln\frac{B}{n_0} - \frac{\partial^2 u_0}{\partial x^2}\right)\phi\right] \tag{46}$$

The plasma frequency of oscillations for ion (electron) is defined as

$$\omega_i(\omega_e) = \sqrt{\frac{n_0 e^2}{M(m)\varepsilon_0}} \tag{47}$$

$$\text{The Poisson's equation } \varepsilon_0\nabla^2\phi = e(n_{e1} - n_{i1}) \tag{48}$$

Using Eqs. (39) and (46) in Eq. (48) gives the perturbed potential in the following form:

$$\frac{\partial^2\phi}{\partial x^2} - k_y^2\phi - \frac{k_y\phi\left(\Omega_z\frac{\partial}{\partial x}\ln\frac{B}{n_0} - \frac{\partial^2 u_0}{\partial x^2}\right)}{(\omega - k_y u_0)\left(1 + \Omega_z^2/\omega_e^2 - \frac{\Omega_z^2\omega_i^2}{\omega_e^2\left(\omega^2 - k_y^2 V_{thi}^2\right)}\right)} = 0 \tag{49}$$

In the case of high frequency of oscillations and in the absence of ion thermal pressure, Eq. (49) turns into Rayleigh's equation of fluid dynamics as below

$$\left(\frac{\partial^2\phi}{\partial x^2} - k_y^2\phi\right) + \frac{\phi k_y}{(\omega - k_y V_y)}\frac{\partial^2 V_y}{\partial x^2} = 0 \tag{50}$$

Here V_y is the flow velocity in the y-direction and ϕ is called the flow function related to $V_y = \vec{\nabla}\phi$. The analytical eigenvalue solution of Eq. (49) is given in Ref. [12].

Resonance condition for the RT instability

From Eq. (49), it is clear that propagating mode may lead to instability if parameter $\Omega_z\frac{\partial}{\partial x}\ln\frac{B}{n_0} - \frac{\partial^2 u_0}{\partial x^2} = 0$ at some point inside the Hall thruster.

7.2 Variations of perturbed potential

The RT Eq. (49) is solved numerically for the perturbed potential ϕ along with the boundary conditions such that $\phi(0) = \phi(d) = 0$. We plot perturbed potential of the instability with magnetic field B, initial drift of the electrons u_0, channel length d, and ion temperature T_i. These parameters can have values as $B = (100 - 250)$ G, $n_0 = 5 \times 10^{17}$–$10^{18}/\text{m}^3$, $T_i = 0{\cdot}1 - 5$ eV, and $u_0 \sim 10^6$ m/s [13, 15].

Figure 5 shows the variation of the perturbed potential with the magnetic field and it has been observed that the potential increases with the increasing magnetic field. These results are consistent with Keidar and Boyd model [28] and that other investigators [13, 14] for the potential of plasma plume. This situation is correspond to the plasma jet enters a transverse magnetic field with a high velocity under the condition that the magnetic field is relatively weak so that only the electrons are magnetized whereas the ions move out of the effect of magnetic field. However, ambipolar (both electrons and ions moving in opposite directions) plasma flow across the magnetic field may require an electric field to appear under the above conditions. Therefore, we can expect the potential to increase across the magnetic field.

The perturbed potential gets increased with the higher value of electron's initial drift velocity (shown in **Figure 6**). Similar behavior of the potential was reported experimentally by King et al. [29] for the potential of plasma plume. Similar results are also reported in Refs. [13, 14]. The enhanced perturbed potential ϕ with the ion temperature is shown in **Figure 7** which is consistent with an experiment [30].

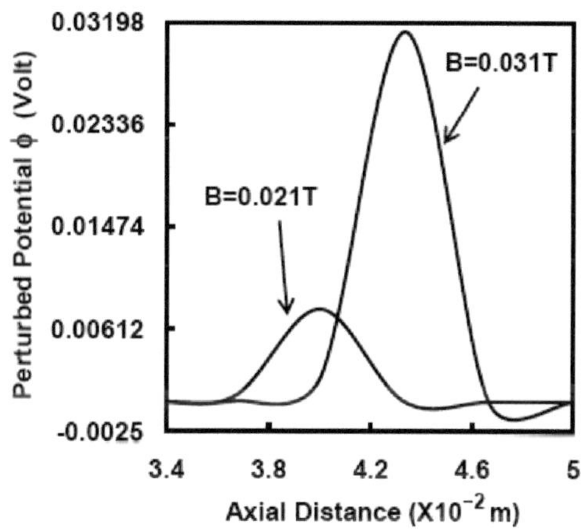

Figure 5.
Effect of magnetic field on the perturbed potential ϕ.

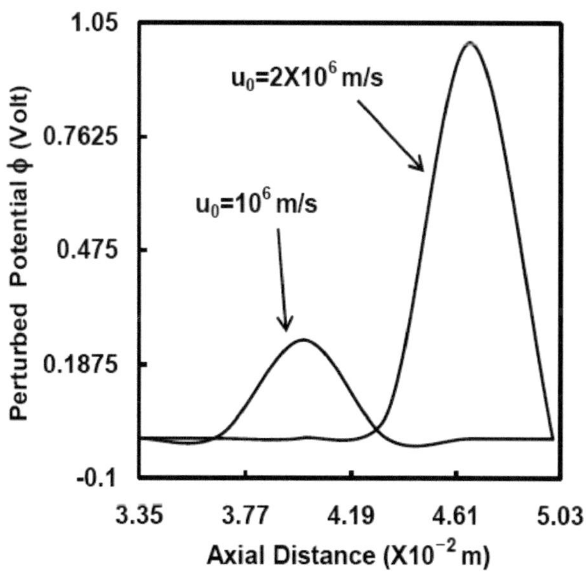

Figure 6.
Dependence of perturbed potential ϕ on the drift velocity of the electrons.

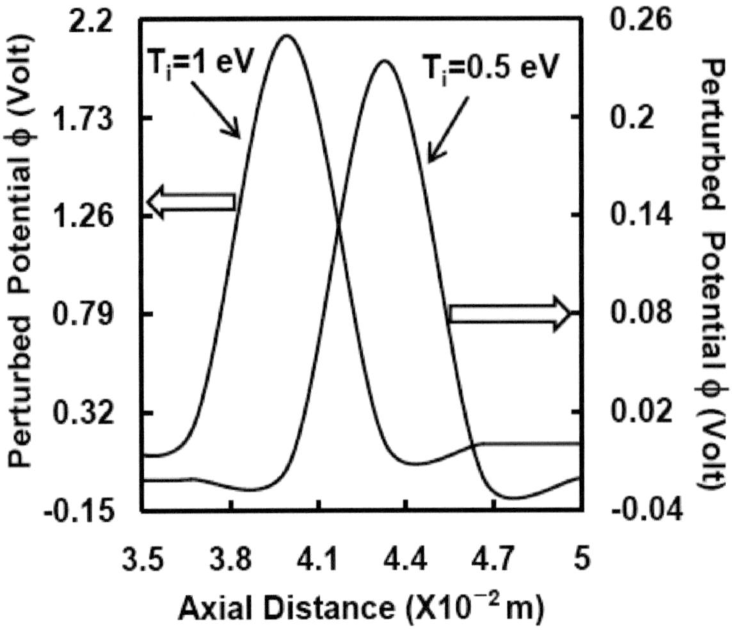

Figure 7.
Variation of perturbed potential ϕ with the ion temperature.

8. Discussions and summary

In conclusion, we can say that short-wavelength perturbations blow up exponentially much more quickly in RTI. The primary source by which this instability is triggered is the gravitational force acting on an inverted density gradient (e.g., a heavy fluid supported by a light fluid). Stable and steady flows may become unstable depending on the ranges of the flow parameters. The instability takes free energy from the mean flow or externally supplied heat and the amplitude of waves grows exponentially. The instabilities exist in all natural and artificial phenomena (in smoke from chimneys, in rivers, in flickering flames) and their effects result in turbulence or random waves. The presence of plasma density and magnetic field gradients is one of the main sources for plasma instabilities in Hall thrusters. It is found that perturbed potential increases with the higher value of electrons' drift velocity, magnetic field, and ion temperature.

Acknowledgements

The University Grants Commission (UGC), New Delhi, India is acknowledged for providing the startup Grant (No. F. 30-356/2017/BSR).

Author details

Sukhmander Singh
Plasma Waves and Electric Propulsion Laboratory, Department of Physics, Central University of Rajasthan, Kishangarh, India

*Address all correspondence to: sukhmandersingh@curaj.ac.in

References

[1] Drazin PG. Introduction to Hydrodynamic Stability (Cambridge Texts in Applied Mathematics). United Kingdom: Cambridge University Press; 2002. DOI: 10.1017/CBO9780511809064

[2] Taylor GI. The instability of liquid surfaces when accelerated in a direction perpendicular to their planes. I. Proceedings of the Royal Society. London. 1950;**A201**:192-196. DOI: 10.1098/rspa.1950.0052

[3] Rayleigh L. On the instability of jets. Proceedings of the London Mathematical Society. 1878;**10**:4-13. DOI: 10.1112/plms/s1-10.1.4

[4] Francois C. Hydrodynamic Instabilities. United Kingdom: Cambridge University Press; 2011. DOI: 10.1017/CBO9780511975172

[5] Qin Y, Lee SM, Pan A, Gösele U, Knez M. Rayleigh-instability-induced metal nanoparticle chains encapsulated in nanotubes produced by atomic layer deposition. Nano Letters. 2008;**8**(1): 114-118

[6] Bychkov V, Marklund M, Modestov M. The Rayleigh–Taylor instability and internal waves in quantum plasmas. Physics Letters A. 2008;**372**(17):3042-3045

[7] Cao J, Ren H, Wu Z, Chu PK. Quantum effects on Rayleigh–Taylor instability in magnetized plasma. Physics of Plasmas. 2008;**15**(1):012110

[8] Khomenko E, Diaz A, De Vicente A, Collados M, Luna M. Rayleigh-Taylor instability in prominences from numerical simulations including partial ionization effects. Astronomy & Astrophysics. 2014;**565**:A45

[9] Diaz AJ, Soler R, Ballester JL. Rayleigh-Taylor instability in partially ionized compressible plasmas. The Astrophysical Journal. 2012;**754**(1):41

[10] Ibrahim EA, Marshall SO. Instability of a liquid jet of parabolic velocity profile. Chemical Engineering Journal. 2000;**76**(1):17-21

[11] Carlyle J, Hillier A. The non-linear growth of the magnetic Rayleigh-Taylor instability. Astronomy & Astrophysics. 2017;**605**:A101

[12] Litvak AA, Fisch NJ. Rayleigh instability in hall thrusters. Physics of Plasmas. 2004;**11**:1379-1382

[13] Singh S, Malik HK. Role of ionization and electron drift velocity profile to Rayleigh instability in a Hall thruster plasma: Cutoff frequency of oscillations. Journal of Applied Physics. 2012;**112**:013307 (1-7)

[14] Malik HK, Singh S. Conditions and growth rate of Rayleigh instability in a Hall thruster under the effect of ion temperature. Physical Review E. 2011; **83**:036406 (1–8)

[15] El Shorbagy KH, Shukla PK. Rayleigh instability in non-uniform multi-ion species magneto-plasmas. Journal of Plasma Physics. 2005;**71** (part 6):747-751

[16] Ali S, Ahmed Z, Mirza AM, Ahmad I. Rayleigh-Taylor/gravitational instability in dense magnetoplasmas. Physics Letters A. 2009;**373**(33):2940-2943

[17] Lighthill J. Waves in Fluids. Cambridge: Cambridge University Press; 2001. ISBN 10: 0521010454; ISBN: 13: 9780521010450

[18] Billingham J, King AC. Wave Motion. Cambridge: Cambridge University Press; 2000

[19] Kaufman HR. Technology of closed drift thrusters. AIAA Journal. 2012;**23**(1):78-86. DOI: 10.2514/3.8874

[20] Ahedo E. Plasmas for space propulsion. Plasma Physics and Controlled Fusion. 2011;**53**(12):124037. DOI: 10.1088/0741-3335/53/12/124037

[21] Jahn RG. Physics of Electric Propulsion. Dover Publications; 2006

[22] Martinez-Sanchez M, Pollard JE. Spacecraft electric propulsion: An overview. Journal Propulsion Power. 1998;**14**(5):688-699

[23] Choueiri EY. Plasma oscillations in hall thrusters. Physics of Plasmas. 2001; **8**(4):1411-1426. DOI: 10.1063/1.1354644

[24] Singh S, Malik HK, Nishida Y. High frequency electromagnetic resistive instability in a Hall thruster under the effect of ionization. Physics of Plasmas. 2013;**20**:102109

[25] Singh S, Malik HK. Growth of low frequency electrostatic and electromagnetic instabilities in a Hall thruster. IEEE Transactions on Plasma Science. 2011;**39**:1910-1918

[26] Singh S, Malik HK. Resistive instabilities in a Hall thruster under the presence of collisions and thermal motion of electrons. The Open Plasma Physics Journal. 2011;**4**:16-23. DOI: 10.2174/1876534301104010016

[27] Malik HK, Singh S. Resistive instability in a Hall plasma discharge under ionization effect. Physics of Plasmas. 2013;**20**:052115

[28] Keidar M, Boyd ID. Effect of a magnetic field on the plasma plume from Hall thrusters. Journal of Applied Physics. 1999, 1999;**86**: 4786-4791

[29] King LB, Gallimore AD, Marrese CM. Transport-property measurements in the plume of an SPT-100 Hall thruster. Journal of Propulsion and Power. 1998;**14**:327-335

[30] Kusamoto D, Mikami K, Komurasaki K, Gallimore AD. Exhaust beam profiles of Hall thruster. Transactions of the Japan Society for Aeronautical and Space. 1998;**40**: 238-247

Chapter 5

Evolutions of Growing Waves in Complex Plasma Medium

Sukhmander Singh

Abstract

The purpose of this chapter to discuss the waves and turbulence (instabilities) supported by dusty plasma. Plasmas support many growing modes and instabilities. Wave phenomena are important in heating plasmas, instabilities, diagnostics, etc. Waves in dusty plasma are governed by the dynamics of electrons, ions and dust particles. Disturbances in solar wind, shocks and magnetospheres are the sources of generation of plasma waves. The strong interest in complex plasma provides us better understanding of physics of dusty universe, solar winds, shocks, magnetospheres, dust control in plasma processing units and surface modifications of materials. The theory of linearization of fluid equation for small oscillation has been introduced. The concept of fine particles in complex plasma and its importance is also explained. The expressions for the growth rate of the instabilities in turbulence plasma have been derived.

Keywords: plasma oscillations, dispersion, turbulence, instabilities, dusty plasma, fine particles, Hall thrusters, resistive plasma, growth rate

1. Introduction to dusty plasma

The presence of fine particles of mass 10^{-10} to 10^{-15} kg and size 1 to 50 micrometer in an electron-ion plasma is called dusty plasma. Dusty plasma also termed complex plasma, plasma crystals, colloidal crystals, fine particle plasma, coulomb crystal or aerosol plasma and has been found in naturally in solar system, planetary rings, interplanetary space, interstellar medium, molecular clouds, circum-stellar clouds, comets, Earth's environments, etc. Manmade plasmas are ordinary flames, dust in fusion devices, rocket exhaust, thermonuclear fusion, Hall thruster, atmospheric aerosols etc. [1–7]. The detail of existence of dusty plasma is given in **Table 1**. Earlier works shows that dust in plasmas has been considered as unwanted constituents and researchers had tried various methods to eliminate dust particles from plasma-processing units. For the moment, Positive aspects of dusty plasmas emerged, dust particles are playing various positive roles in plasma processing devices. The dust particles experience different forces in plasma. The Gravitational force, drag force, electromagnetic forces, polarization force and radiation pressure [1–8].

2. Physical processes in dusty plasma

There are many circumstances when astrophysical plasma and dust particles are found to coexist together. The study of dusty plasmas systems has an exciting

Cosmic dusty plasmas	Dusty plasmas in the solar system	Dusty plasmas on the earth	Man-made dusty plasmas
Solar nebulae	Cometary tails and comae	Ordinary flames	Rocket exhaust
Planetary nebulae	Planetary ring Saturn's rings	Atmospheric aerosols	Dust on surfaces of space vehicle
Supernova shells	Dust streams ejected from Jupiter	charged snow	Microelectronic fabrication
Interplanetary medium	Zodiacal light	lightning on volcanoes	Dust in fusion devices
Molecular clouds	Cometary tails and comae		Thermonuclear fireballs
Circumsolar rings			Dust precipitators used to remove pollution from
Asteroids			

Table 1.
Classification of dusty plasmas.

properties which has attracted researchers over the world. These fine particles acquire some charges from the electrons to get charged. Moreover, in ordinary plasma, the charge considered to be constant on each particle, whereas, the charge on the dust particle varies with time and position [9, 10]. The charge on the dust particle generally depends on the type of dust grain, the surface properties of dust grain, the dust dynamics, the temperature, density of plasma and the wave motion in the medium. The plasma environments around these particles determine the nature of the charge (positive or negative) of these dusty plasmas. Although, most of the cases, these charged dust particles are negatively charged through different charging process. Their electric charge is determined by the size and composition of the grains [9, 10]. The fact that the frequencies associated with dust particles are smaller than those with electrons -ions and presences of fine particles modifies the dynamics of plasma motions and give rise to new types of propagating modes.
For Dust acoustic waves, where ions and electrons are supposed to be inertia less pressure as gradient is balanced by the electric force, leading to Boltzmann electron and ion number density perturbations, whereas the mass of the dust play an important role in dust dynamics. In the dust acoustic wave the inertia is provided by the massive dust particles and the electrons and ions provide the restoring force. The effect of dust is to increase the phase velocity of the ion acoustic waves. This can be interpreted formally as an increase in the effective electron temperature which has important consequences for wave excitation. The dusty plasma also has a trend to oscillate at its plasma frequency [9, 10].

3. Parameters of dusty plasma

Dusty plasma and ordinary (electron-ion) plasma are different from each other due to the charge to mass ratio difference.

3.1 Dust plasma frequency

The electron and ion plasma frequency is much greater than the dust plasma frequency and it is defined as

$$\omega_{pd} = \sqrt{\frac{Z^2 e^2 n_{d0}}{\varepsilon_o m_d}} < < \omega_{pi}, \omega_{pe} \tag{1}$$

3.2 Gyro frequency in dusty plasma

When charged dust grain/particle executes a spiral motion about the magnetic lines of force, then dust particle moves perpendicular to the magnetic field with Gyro frequency of the dust particle. The centrifugal force is balanced by the Lorentz force. In mathematically,

$$\frac{m_d v^2}{r_d} = eZvB \tag{2}$$

The radius of gyration is

$$r_d = \frac{m_d v}{eZB} = \frac{v}{\Omega_{cd}} \tag{3}$$

where, $\Omega_{cd} = \frac{eZB}{m_d}$, is called the dust cyclotron frequency.

3.3 Macroscopic neutrality

The quasi-neutrality condition is obtained for the negatively charged dusty plasma by $n_{e0} = n_{i0} + Zn_{d0}$, here n_{d0} is equilibrium dust particle density and Z is the electric charge number on the dust particles.

3.4 Strongly vs. weakly coupled dusty plasma (Coulomb correlation parameter)

The property of dust particles in the plasma is expressed by coupling parameter Γ. It is the ratio of the interparticle Coulomb potential energy to the thermal energy of the particles. When the value of Γ exceeds unity, the species are termed to be strongly coupled otherwise weakly coupled dusty plasma. if r_d is the average interparticle separation between particles, then coupling parameter

$$\Gamma = \frac{e^2 Z^2}{4\pi\varepsilon_0 r_d k_B T_d} \tag{4}$$

The interparticle separation can be found out by the relation $r_d = \left(\frac{4\pi n_d}{3}\right)^{-\frac{1}{3}}$. For the typical values of $Ze = 5000e$, $k_B T_d = 0.05$ eV, $n_d = 10^{10}$ m^{-3}, the coupling parameter comes out to be 1500. It is also experienced that, when $\Gamma \sim 170$, the dust particles are found in arranged fashion and said to be Coulomb crystals.

4. Applications of ordinary plasma

Plasma technology is safe, less costly and playing important roles in every fields of daily life. Some of these applications are discussed in **Table 2**.

Fields	Applications
Telecommunication	The Global Positioning System (GPS) use ionosphere's plasma layer to reflect the signal transmitted by GPS satellite for further communication usage.
Sterilization	To sterilize the surgical equipments, which are directly connected with patient's immune system, where cleanliness is difficult
Medical treatment	Plasma treatment is contact-free, painless hardly damage tissue.
In dentistry	Plasmas treatment are used inside the root canal to kill the bacteria
Pollution controlling	Plasma technology is used to control gaseous and solid pollutions.
Water Purification	for destroying viruses and bacteria in a water, Ozone (O3) generated by plasma technology is more effective and less costly at large scale than existing chlorination method
Etching and cleaning of materials	To removes contaminants and thin layers of the substratum by bombarding with the plasma species which break the covalent bonds. It is also used to control the weight of the exposed substrate.
fusion research	Plasma is used to achieve high temperature to run the controlled thermonuclear fusion reactors
nanotechnology	Plasma discharges are helpful in growing the nanoparticles for nano world.

Table 2.
Applications of plasma in different fields.

4.1 Applications of dusty plasma

The presence of dust particles in a system also has positive impacts and has many applications in nanotechnology to synthesize the desired shape and size of the particles by controlling the dynamics of charged dust grains. Surface properties of the exposed materials could be improved by coating with plasma enhanced chemical vapor deposition method. Methane plasma is used to synthesize productive Carbon nanostructures which have like high hardness and chemical inertia. Dusty plasmas are also used for the fabrication of semiconductor chips, solar cells and flat panel displays.

5. Current status of the research

As we know that plasma support electrostatic as well as electromagnetic waves because of the motions of the charged particle. Studies of these waves provide the useful information about the state of the system. The resonance frequencies of plasmas waves can be used as diagnostics tool to characterize the plasma parameters. Plasma waves are generated for acceleration of energetic particles and heating plasmas. The exponential growing waves and modes in plasma removes the free energy from the system and permit the system to become unstable. The study of dusty plasma has gained interest in the last few decades due to its observations [1–10] and applications in the space and laboratory [1–10]. Many authors studied the linear and nonlinear electrostatic wave in the presence and absence of the external magnetic field [11–13]. Sharma and Sugawa studied the effect of ion beam in dusty plasma on ion cyclotron wave instability [13]. The presence of the charged dust grains in the plasma modifies the collective behavior of a plasma and excites the new modes [12–14].

The present charged dust particles introduces dust acoustic and dust ion acoustic waves in the plasma after altering the dynamics of electrostatic and electromagnetic waves of ordinary plasma [11–14]. The charged dust grain also introduces growing and damped modes. Tribeche and Zerguini studied the dust ion-acoustic waves in collisional dusty plasma [15]. Rao et al. [16] predicted the existence of dust-acoustic wave in an unmagnetized plasma that has inertial dust and Maxwellian distributed electrons and ions. Shukla and Silin [17] showed the existence of dust-ion acoustic wave in a plasma. Barkan experimentally investigated that negatively charged dust grains enhances the growth rate of the electrostatic ion cyclotron instability [18]. Akhtar et al. [19] studied the dust-acoustic solitary waves in the presence of hot and cold dust grains. The existence of dust-acoustic wave and dust ion acoustic wave has been confirmed by many investigators in a laboratory experiments [18–20]. Ali [21] reported the electrostatic potential due to a test-charge particle in a positive dusty plasma. Bhukhari et al. derived generalized dielectric response function for twisted electrostatic waves in unmagnetized dusty plasmas [22]. Mendonça et al. showed that a modified Jeans instability lead to the formation of photonbubble in a dusty plasma which in turn form two different kinds of dust density perturbations [23]. Pandey and Vranjes predicted that growth rate of the instability is proportional to the whistler frequency in a magnetized dusty plasma [24].

6. Plasma model and basic equations

Phase and group velocity can be calculated by finding the relation between ω and k. This relation $\omega = \omega(k)$, is called the dispersion relation and contains all the physical parameters of the given medium in which wave propagates. If the frequency has an imaginary part, that indicates an instability. Plasma instability involves some growing modes, whose amplitude increases exponentially. In other words instability represents the ability of the plasma to escape from a configuration of fields [8, 9].

6.1 Electrostatic and electromagnetic waves in ordinary (electron-ion) plasma

Charged particles in a plasmas couples to electromagnetic field. Because of this effect various kinds of waves are formed in plasmas. Plasma waves are electrostatic or electromagnetic based on perturbed (oscillated) magnetic field. If there is a perturbed magnetic field ($\vec{B}_1 \neq 0$), plasma support electromagnetic waves. If the oscillating magnetic field associated with the wave is absent ($\vec{B}_1 = 0$), then only electrostatic waves are supported by plasma. In addition, Electrostatic waves may have longitudinal and transverse component depending on the direction of propagation with the perturbed electric field [8, 9].

A thermal unmagnetized plasma support many modes as discussed by Tonks and Langmuir in 1929. One is transverse waves in a plasma have dielectric constant $\varepsilon_r(\omega) = \frac{\varepsilon(\omega)}{\varepsilon_0} = \left(1 - \frac{\omega_{pe}{}^2}{\omega^2}\right)$. In case of a lower frequency wave ($\omega < \omega_{pe}$), the dielectric constant would be negative. It turn out that if refractive index become imaginary, then waves cannot propagate but are damped (absorbed). Therefore plasma behaves like a waveguide with propagation and cut-off regions depending on the range of frequencies [8, 9].

The electron plasma wave and ion acoustic wave are the examples of electrostatic longitudinal modes, that is particle oscillate parallel to the direction of wave propagation. Ion acoustic waves are electrostatic waves, when both ions and electrons are allowed to oscillate in the wave-field. IA waves are low frequency longitudinal wave and we can use the plasma approximation, $n_{e1} \approx n_{i1} \approx n_0$.

The electron plasma wave (Langmuir mode) satisfy the dispersion relation $\omega(k) = \sqrt{\omega_{pe}{}^2 + 3k^2 \frac{k_B T_e}{m_e}}$, whereas the ion acoustic mode satisfy the dispersion relation $\omega(k) = \frac{k}{\sqrt{1+\lambda_{De}{}^2 k^2}} \frac{k_B T_e}{m_i}$, here λ_{De} is the Debye length of electron [8, 9]. The propagation of electromagnetic waves in the unmagnetized plasma yield the dispersion relation $\omega(k) = \sqrt{\omega_{pe}{}^2 + k^2 c^2}$.

7. Theoretical formulation for the studies of waves in dusty plasma

We consider a unmagnetized collisionless plasma consisting of electrons, ions and dust particles. Here we use the fluid equations and Maxwell's equations to derive the dispersion relations in dusty plasma corresponding to ordinary electron –ion plasma. We denote $\vec{v}_\alpha$ and n_α are the plasma velocity and density of the different species $(\alpha = e, i, d)$ having mass m_α, temperature T_α in electron-volt. We write the equations of continuity and equation of motion of particles to derive the dispersion relation. Then the equations of motion governing the plasma can be written as

$$\frac{\partial n_\alpha}{\partial t} + \vec{\nabla} \cdot \left(\vec{v}_\alpha n_\alpha\right) = 0 \tag{5}$$

$$\frac{d\vec{v}_\alpha}{dt} = \left\{\frac{\partial}{\partial t} + \left(\vec{v}_\alpha \cdot \vec{\nabla}\right)\right\}\vec{v}_\alpha = \frac{Q}{m_\alpha}\left(\vec{E} + \vec{v}_\alpha \times \vec{B}\right) - \frac{T_\alpha \vec{\nabla} n_\alpha}{m_\alpha n_\alpha} \tag{6}$$

If we define thermal velocities $V_{T\alpha} = \sqrt{\frac{T_\alpha}{m_\alpha}}$.

In the above equation, the derivative $\frac{d\vec{v}_\alpha}{dt}$ is called the convective derivative. $\frac{d\vec{v}_\alpha}{dt}$ can be viewed as the time derivative of $\vec{v}_\alpha$ taken in a "fluid" frame of reference moving with a velocity of $\vec{v}_\alpha$ relative to a rest frame. $\frac{\partial \vec{v}_\alpha}{\partial t}$ represents the rate of change of $\vec{v}_\alpha$ at a fixed point in space and $\left(\vec{v}_\alpha \cdot \vec{\nabla}\right)\vec{v}_\alpha$ represents the change of $\vec{v}_\alpha$ measured by an observer moving in the fluid frame into a region where $\vec{v}_\alpha$ is inhomogeneous.

7.1 Linearization of fluid equations

We consider the perturbed density $n_{\alpha 1}$ and velocity $\vec{v}_{\alpha 1}$ indicated by subscript 1 along with their unperturbed density $n_{\alpha 0}$ and velocity $u_{\alpha 0}$. The unperturbed electric field (magnetic field) as $\vec{E}_0$ $(\vec{B}_0)$ and the perturbed value of the electric field (magnetic field) is taken as $\vec{E}_1$ $(\vec{B}_1)$. To linearize all the equations, let us write $n_\alpha = n_{\alpha 0} + n_{\alpha 1}$, $\vec{v}_\alpha = \vec{v}_{\alpha 1} + \vec{v}_{\alpha 0}$ and $\vec{E} = \vec{E}_1 + \vec{E}_0$. If the amplitude is chosen to be much smaller than the wavelength of the instability, the equations of motion can be linearized. The perturbed quantities $f_{\alpha 1}$ are much smaller than their unperturbed values $f_{\alpha 0}$, that is $f_{\alpha 1} << f_{\alpha 0}$. If $\vec{v}_{\alpha 0}$ and $n_{\alpha 0}$ are constant, the terms $\left(\vec{v}_{\alpha 0} \cdot \vec{\nabla}\right) n_{\alpha 0}$, $n_{\alpha 0}\left(\vec{\nabla} \cdot \vec{v}_{\alpha 0}\right)$ and $n_{\alpha 1}\left(\vec{\nabla} \cdot \vec{v}_{\alpha 0}\right)$ are equal to be zero. Further the terms $\left(\vec{v}_{\alpha 1} \cdot \vec{\nabla}\right) n_{\alpha 1}$, and $n_{\alpha 1}\left(\vec{\nabla} \cdot \vec{v}_{\alpha 1}\right)$ are neglected as they are quadratic in perturbation. The linearized form of the fluid equations can be written as

$$\frac{\partial n_{\alpha 1}}{\partial t} + n_{\alpha 0}\vec{\nabla}\cdot(v_{\alpha 1}) + v_{\alpha 0}.\vec{\nabla}n_{\alpha 1} = 0 \tag{7}$$

$$\frac{\partial v_{\alpha_1}}{\partial t} + v_{\alpha 0}\left(\vec{\nabla}\cdot v_{\alpha 1}\right) + \frac{V^2{}_{th\alpha}}{n_{\alpha_0}}\vec{\nabla}n_{\alpha 1} = \frac{Q}{m_\alpha}\left(-\vec{\nabla}\varphi_1 + \vec{v}_{\alpha 1}\times\vec{B}_0\right) \tag{8}$$

$$\varepsilon_0\nabla^2\varphi_1 = \rho = e(n_{e1} + Zn_{d1} - n_{i1}) \tag{9}$$

Let us define $\delta = \frac{n_{d0}}{n_{i0}}$ is the relative dust density, then quasi-neutrality condition follow

$$\frac{n_{e0}}{n_{i0}} = Z\delta + 1 \tag{10}$$

Thus, the assumptions of small oscillation give a set of linear equations.

7.2 Dust-acoustic waves (DAW)

The DAW is an electrostatic wave generated in dusty plasma, where inertia is provided by the dust grains. It is same to the ion-acoustic wave in general plasma, where inertia is provided by the ions. The frequency ω_{pd} of dust acoustic wave is very low due to the high dust mass than the ion (electron) plasma frequency $(\omega_{pi}, \omega_{pe})$. That why dusty plasma supports low frequencies waves. In mathematically, $\omega_{pd} = \sqrt{\frac{e^2Z^2n_{d0}}{\varepsilon_o m_d}} < < \omega_{pi}, \omega_{pe}$, where n_{d0} is the equilibrium dust density.

Let us consider a situation, when dust density is get disturbed. This change will alter the charge on the dust particles and results to an enhancing negative space charge due to the process of negative dust charging. This total space charge density of dust $\rho_d = eZn_{d0}$ is shielded by the surrounding plasma ions and electrons. Therefore an electric field is generated due to the space charge by the fluctuations of dust charge density. This oscillating electric field imparts the force on the dust particle, which further pushes the fluctuations in the direction of the electric field and thus the wave propagates.

We consider that fluctuations are plane wave, which propagating inside the dusty plasma having the form $f = f_o \exp\left\{i\left(\vec{k}\cdot\vec{r} - \omega t\right)\right\}$. Them the time derivative $(\partial/\partial t)$ can be replaced by $-i\omega$ and the gradient $\vec{\nabla}$ by ik. Here $f_1 \equiv n_{\alpha 1}$, $\vec{v}_{\alpha 1}$, $\vec{E}_1$, $\vec{B}_1$. The electrons and ions are assumed to inertia less as compared with mass of the dust grains and should have a Boltzmann distribution, namely

$$n_e = n_{e0}\exp\left(\frac{e\phi_1}{T_e}\right) \cong n_{e0}\left(1 + \frac{e\phi_1}{T_e}\right) \tag{11}$$

$$n_i = n_{i0}\exp\left(-\frac{e\phi_1}{T_i}\right) \cong n_{i0}\left(1 - \frac{e\phi_1}{T_i}\right) \tag{12}$$

Here, n_{e0} and n_{i0} denote the unperturbed values of the electron and ion density respectively. Let us limit that, all the unperturbed velocities are zero, then equation of continuity and equations of motion follows.

The above three equations can be written as

$$-i\omega\, n_{d1} + ikn_{d0}v_{d1} = 0 \tag{13}$$

$$-i\omega\, v_{d1} + ik\frac{V^2{}_{thd}}{n_{d0}}n_{d1} = \frac{Ze}{m_d}ik\phi_1 \tag{14}$$

$$-k^2\phi_1 = \frac{e}{\varepsilon_0}(n_{e0} + n_{e1} + Zn_{d0} + Zn_{d1} - n_{i0} - n_{i1}) \tag{15}$$

Eqs. (13) and (14) gives

$$n_{d1} = \frac{k^2 n_{d0}\phi_1 Ze}{m_d\left(k^2 V^2{}_{thd} - \omega^2\right)} \tag{16}$$

After substituting into Poisson's equation, we obtain

$$-k^2\phi_1 = \frac{e}{\varepsilon_0}(n_{e0} + Zn_{d0} - n_{i0}) + \frac{e}{\varepsilon_0}\left\{n_{e0}\left(1 + \frac{e\phi_1}{T_e}\right) - n_{i0}\left(1 - \frac{e\phi_1}{T_i}\right)\right\} + Z\frac{e}{\varepsilon_0}n_{d1} \tag{17}$$

The first term reduces to zero under the quasi-neutrality condition ($n_{i0} = n_{e0} + Zn_{d0}$). Let, the relative dust density is defined by $\delta = n_{d0}/n_{i0}$, then we get

$$-k^2\phi_1 = \frac{e^2 n_{i0}}{\varepsilon_0 T_{i0}}\left(1 + \frac{T_i}{T_e}(1 - \delta Z_d)\right)\phi_1 + Z\frac{e}{\varepsilon_0} \times \frac{k^2 n_{d0}\phi_1 Ze}{m_d\left(k^2 V^2{}_{thd} - \omega^2\right)} \tag{18}$$

After simplification for the nontrivial solution, we readily obtain

$$\omega^2 = k^2 V^2{}_{thd} + \frac{\omega_{pd}{}^2}{1 + \frac{1}{\lambda_{Di}{}^2 k^2}\left(1 + \frac{T_i}{T_e}(1 - \delta Z_d)\right)} \tag{19}$$

the dispersion relation of the DAW shows depends on dust density, temperature of electron, ion and dust. It also shows depends on the inertia of electron and ion.

7.2.1 Limiting cases

In the limit of cold dust ($T_d = 0$) and cold ions ($T_i < < T_e$). Then, the dispersion relation simplifies into the dispersion relation of ion-acoustic wave in an ordinary plasma

$$\omega = \frac{k\lambda_{Di}\omega_{pd}}{\sqrt{1 + \lambda_{Di}{}^2 k^2}}, \tag{20}$$

which is the same as for the ion-acoustic wave in classical plasma.

7.2.2 Behavior at low wave length

For small wave numbers $k^2\lambda_{D,i}^2 < < 1$ the wave is acoustic $\omega = kC_{DAW}$ with the dust-acoustic wave speed

$$C_{DAW} = \sqrt{\varepsilon Z_d^2 \frac{kT_i}{m_d}} \tag{21}$$

7.2.3 Behavior at high wave length

For large wave numbers $k^2\lambda_{D,i}^2 > > 1$, the wave is not propagating and just oscillates at the dust plasma frequency.

7.3 Dust ion acoustic wave (DIAW)

In the previous expression of ion-acoustic wave, the wave speed depends on ion temperature and on the mass of the dust. In the DIAW, the dust particles are supposed to immobile. We write the equation of motion, continuity and Poisson's equation

$$\frac{\partial n_i}{\partial t} + \vec{\nabla} \cdot (v_i n_i) = 0 \tag{22}$$

$$\frac{\partial v_i}{\partial t} + v_i \frac{\partial v_i}{\partial x} = \frac{e}{m_i} \frac{\partial \phi}{\partial x} \tag{23}$$

$$\frac{\partial^2 \phi}{\partial x^2} = -\frac{e}{\varepsilon_0}(n_i - n_e) \tag{24}$$

The Poisson's equation contains the perturbed electron and ion densities. The electrons are treated as Boltzmann distributed as follow

$$n_e = n_{e0} \exp\left(\frac{e\varphi}{kT_e}\right) \tag{25}$$

The only place, where the dust properties enter is the quasi-neutrality condition

$$n_{i0} = n_{e0} + Z_d n_{d0} \tag{26}$$

Eqs. (24), (25) and (26) gives the dispersion relation for the DIAW as

$$\omega^2 = \frac{\omega_{pi}^2 k^2 \lambda_{D,e}^2}{1 + k^2 \lambda_{D,e}^2} = \left(\frac{n_{i0}}{n_{e0}}\right) \frac{kT_e}{m_i} \frac{k^2}{1 + k^2 \lambda_{D,i}^2} \tag{27}$$

Using Eq. (25)

$$\omega^2 = \frac{\omega_{pi}^2 k^2 \lambda_{D,e}^2}{1 + k^2 \lambda_{D,e}^2} = \left(1 + \frac{Zn_{d0}}{n_{e0}}\right) \frac{kT_e}{m_i} \frac{k^2}{1 + k^2 \lambda_{D,i}^2} \tag{28}$$

It is clear from Eq. (27), that phase speed of the DIAW is increase as dust charge density increases. But the electron density has opposite effect on the speed of the DIAW.

8. Dissipative turbulence/instabilities in Hall thruster plasma

The section is devoted to the existing instabilities in a Hall thruster plasma. The principle of thrusters is the ionization of a Noble gas (propellant) in a crossed filed discharge channel. The accelerated heavy ions of inert gas are used to generate a thrust by the use of electrostatic forces. Xenon is used as an ion thruster propellant because of its low reactivity with the chamber and high molecular weight [25–33]. These types of devices support many waves and instabilities because of the turbulence nature of the plasma. These instabilities affect the performance and the efficiency of the device. In order to control these instabilities and further consequences, it has become necessary to study the growth rate of these instabilities. In a Hall thruster, the electrons experiences force along the azimuthal direction because of $\vec{E} \times \vec{B}$ drift. The collision momentum transfer frequency (v) between the electrons and neutral atoms are also taken into account to see the resistive effects in the plasma. Since ions do not feel magnetic field because of their larger larmor radius compared to length of the device. There equation of motion for ions can be written as

$$M\left\{\frac{\partial}{\partial t}+\left(\vec{v}_i\cdot\vec{\nabla}\right)\right\}\vec{v}_i=e\vec{E} \tag{29}$$

Motion of electrons under the electric and magnetic fields

$$mn_e\left\{\frac{\partial}{\partial t}+\left(\vec{v}_e\cdot\vec{\nabla}\right)+v\right\}\vec{v}_e=-en_e\left(\vec{E}+\vec{v}_e\times\vec{B}\right)-\vec{\nabla}p_e \tag{30}$$

8.1 Linearization of fluid equations

Let us denote the perturbed densities for ions and electrons as n_{i1} and n_{e1} velocities as $\vec{v}_{i1}$ and $\vec{v}_{e1}$ respectively. The unperturbed velocities v_0 and u_0 are taken in the x- and y-direction respectively. The amplitude of oscillations of the perturbed densities are taken small enough. The linearized form of Eq. (29) and Eq. (30) are written as

$$M\left(\frac{\partial}{\partial t}+v_0\frac{\partial}{\partial x}\right)\vec{v}_{i1}=-e\vec{\nabla}\phi_1 \tag{31}$$

$$m\left(\frac{\partial}{\partial t}+u_0\frac{\partial}{\partial y}+v\right)\vec{v}_{e1}=e\left(\vec{\nabla}\phi_1-\vec{v}_{e1}\times\vec{B}_0\right)-\frac{T_e\vec{\nabla}n_{e1}}{n_0} \tag{32}$$

The continuity equations of electrons and ions can be linearized as below

$$\left(\frac{\partial}{\partial t}+v_0\frac{\partial}{\partial x}\right)n_{i1}+n_0\left(\vec{\nabla}\cdot\vec{v}_{i1}\right)=0 \tag{33}$$

$$\left(\frac{\partial}{\partial t}+u_0\frac{\partial}{\partial y}\right)n_{e1}+n_0\left(\vec{\nabla}\cdot\vec{v}_{e1}\right)=0 \tag{34}$$

Fourier analysis: We seek the sinusoidal solution of the above equations, therefore the perturbed quantities are taken as $f_1 \sim f_0 \exp\left(i\omega t - i\vec{k}\cdot\vec{r}\right)$. Them the time derivative $(\partial/\partial t)$ can be replaced by $i\omega$ and the gradient $\vec{\nabla}$ by ik, here $f_1 \equiv n_{i1}$, n_{e1}, ϕ_1, $\vec{v}_{i1}$, $\vec{v}_{e1}$, $\vec{E}_1$ together with ω as the frequency of oscillations and $\vec{k}$ as the propagation vector.

By using Fourier analysis from Eq. (31)–(34), the perturbed ion and electron densities are given as follows,

$$n_{i1}=\frac{ek^2n_0\phi_1}{M(\omega-k_xv_0)^2} \tag{35}$$

$$n_{e1}=\frac{en_0(\omega-k_yu_0-iv)k^2\phi_1}{m\Omega^2(\omega-k_yu_0)+m(\omega-k_yu_0-iv)k^2{V_{th}}^2} \tag{36}$$

The expression for the electron density n_{e1} is derived under the assumptions that $\Omega >> \omega$, k_yu_0 and v in view of the oscillations observed in Hall thrusters.

8.2 Dispersion equation and growth rate of electrostatic oscillations

Finally, we use the expressions for the perturbed ion density n_{i1} and electron density n_{e1} in the Poisson's equation $\varepsilon_0 \nabla^2 \phi = e(n_{e1} - n_{i1})$ in order to obtain

$$-k^2\phi_1 = \frac{\omega_e{}^2 \hat{\omega} k^2 \phi_1}{\Omega^2(\omega - k_y u_0) + \hat{\omega} k^2 V_{th}{}^2} - \frac{\omega_i{}^2 k^2 \phi_1}{(\omega - k_x v_0)^2} \tag{37}$$

For the nontrivial solution of the above equation, the perturbed potential $\phi_1 \neq 0$, we have from Eq. (37)

$$\frac{\omega_e{}^2 \hat{\omega}}{\Omega^2(\omega - k_y u_0) + \hat{\omega} k^2 V_{thE}{}^2} + \frac{(\omega - k_x v_0)^2 - \omega_i{}^2}{(\omega - k_x v_0)^2} = 0 \tag{38}$$

This is the dispersion relation that governs the electrostatic waves in the Hall thruster's channel. In the above equations, we introduced parameter $\omega_{e(i)} = \sqrt{\frac{e^2 n_0}{m(M)\varepsilon_0}}$, $\Omega = \frac{eB_0}{m}$ and $V_{th} = \sqrt{\frac{Y_e T_e}{m}}$.

After simplification of Eq. (38) we obtain

$$\begin{aligned}
&\omega^3(\omega_e{}^2 + \Omega^2 + k^2 V_{th}{}^2) - \omega^2\left[(\omega_e{}^2 + \Omega^2 + k^2 V_{th}{}^2)(k_y u_0 + 2k_x v_0) + iv(\omega_e{}^2 + k^2 V_{th}{}^2)\right] \\
&+\omega\left[k_x{}^2 v_0{}^2 \omega_e{}^2 + 2k_x v_0 (k_y u_0 \omega_e{}^2 + iv\omega_e{}^2 + ivk^2 V_{th}{}^2) + (k^2 V_{th}{}^2 + \Omega^2)(2k_x v_0 k_y u_0 + k_x{}^2 v_0{}^2 - \omega_i{}^2)\right] \\
&-(k_x{}^2 v_0{}^2 - \omega_i{}^2)\left[k_y u_0(\Omega^2 + k^2 V_{th}{}^2) + ivk^2 V_{th}{}^2\right] - \omega_e{}^2 k_x{}^2 v_0{}^2 (k_y u_0 + iv) = 0
\end{aligned} \tag{39}$$

This is the dispersion equation that governs the electrostatic waves in the Hall thruster's channel. It is clear from the above equation that Hall thruster support different waves and instabilities which satisfies the dispersion relation (39).

8.3 Results and discussion

To estimate the growth rates of the instability, we numerically solve Eq. (39) by giving typical values of all parameters used for the thruster [25–32]. Therefore, for investigating the growths of the waves, we plot the negative imaginary parts of the complex roots (correspond to the instabilities) in **Figures 1–3**.

It is found that the growth of the wave is enhanced for the larger values of the wave number of the oscillations as shown in **Figure 1**. In others words, the oscillations of larger wavelengths are stable. The findings are consistent to the results predicted by Kapulkin et al. [34]. In **Figure 2**, the variation of growth of the wave with the momentum transfer collision frequency is shown and it has been depicted that instability grow at faster rates in the presence of more electron collisions. This is mainly due to the resistive coupling with electrons' drift's in the presence of more collisions. The similar results are also predicted by Fernandez et al. [35] in the simulation studies of dissipative instability which shows that the growth of the instability is directly proportional to the square root of the collision frequency.

In **Figure 3** it is noted that the wave grows faster if the electrons carry higher temperature. The present finding matched with the same results observed by other investigators [36, 37] of dissipative instability in an plasma.

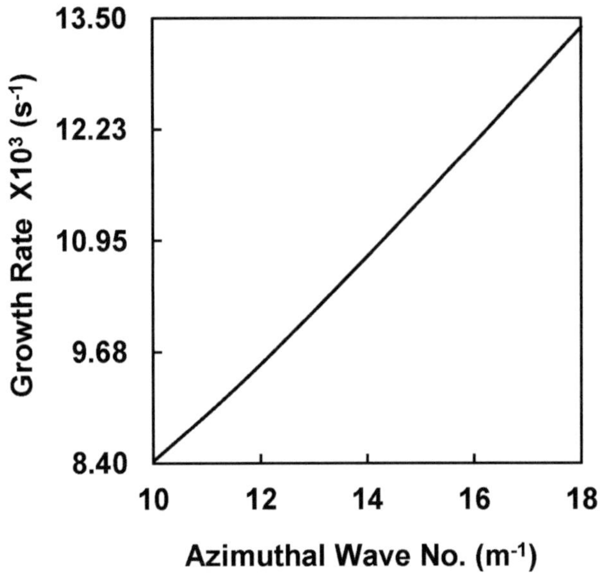

Figure 1.
Growth rate versus azimuthal wave number, with parameters of Hall plasma thrusters as $B_0 \sim 100 - 200\ G$, $n_0 \sim 5 \times 10^{17}\ 10^{18}\ m^3$, $T_e = 10 - 15\ eV$, $u_0 \sim 10^6\ m/s$, $v \sim 10^6\ /s$ *and* $v_0 \sim 2 \times 10^4 - 5 \times 10^4\ m/s$.

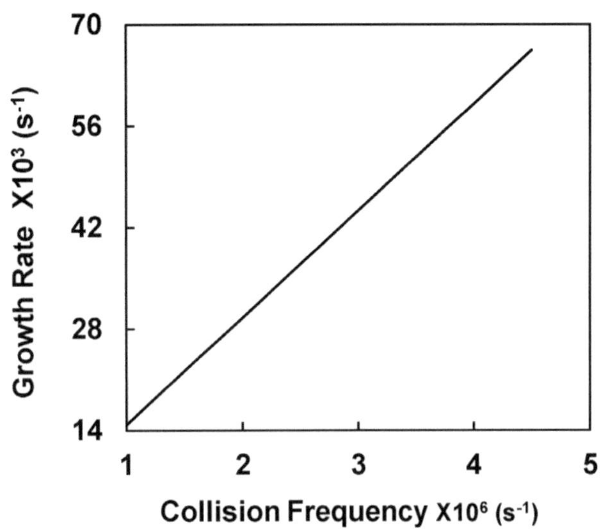

Figure 2.
Growth rate versus collision frequency.

8.4 Conclusion

In summary, we can say that dusty plasma physics has vital role in novel material processing and diagnostics tools. The nanoparticles of desired shape can be synthesized by controlling the dynamics of charged particles in the semiconductor industry. For example, the rotation of the dust particles can extract the electron flux in the magnetron sputtering unit. Thus, dusty plasma is a remarkable field in all areas of natural sciences. Though, the elimination of dust particles in the semiconductor industry is still a main alarm. The advanced development tools to learning

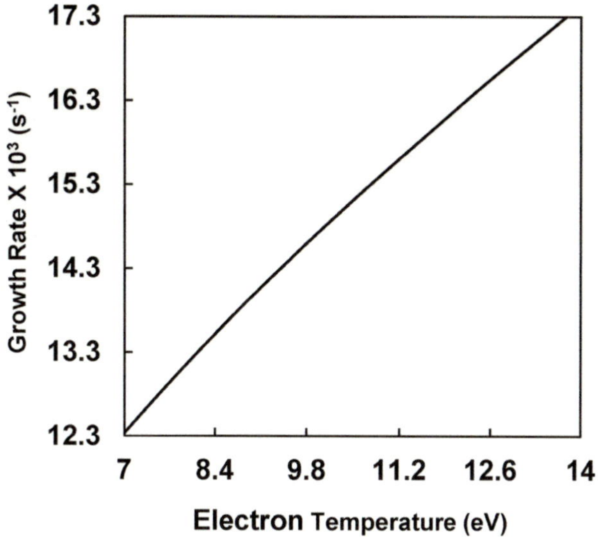

Figure 3.
Variation of growth rate with electron temperature.

dusty plasma will lead to additional discoveries about the astrophysical events and new scientific findings. The different dispersion relations are derived to see the behavior of the wave with wave number in complex plasma. The theory of linearization has been used for the smaller amplitude of the oscillations to derive the perturbed quantities. In the last section, the growth rate of dissipative instability has been depicted in a Hall thruster turbulence plasma. The instability grow faster with the collision frequency, azimuthal wave number and the electron temperature.

Acknowledgements

The University Grants Commission (UGC), New Delhi, India is thankfully acknowledged for providing the startup Grant (No. F. 30-356/2017/BSR).

Author details

Sukhmander Singh
Plasma Waves and Electric Propulsion Laboratory, Department of Physics, Central University of Rajasthan, Ajmer, Kishangarh, India

*Address all correspondence to: sukhmandersingh@curaj.ac.in

References

[1] Shukla PK, Mamun AA. Series in Plasma Physics: Introduction to Dusty Plasma Physics. Bristol: Taylor & Francis Group, CRC Press; 2001

[2] Kersten H, Thieme G, Fröhlich M, Bojic D, Tung D H, Quaas M, Wulff H, Hippler R. Complex (dusty) plasmas: Examples for applications and observation of magnetron induced phenomena 2005 Pure and Applied Chemistry. 77 415. DOI: 10.1351/pac200577020415

[3] Merlino RL. Dusty plasmas and applications in space and industry. Plasma Physics Applied. 2006. ISBN: 81-7895-230-0

[4] Bonitz M, Henning C, Block D. Complex plasmas: A laboratory for strong correlations. Reports on Progress in Physics 2010;**73**(6):066501. DOI: 10.1088/0034-4885/73/6/066501

[5] Dusty Plasmas in the Laboratory, Industry, and space physics. Today. 2004;**57**(7):32. DOI: 10.1063/1.1784300

[6] Kersten H, Wolter M. Complex (dusty) plasmas: Application in material processing and tools for plasma diagnostics. Introduction to Complex Plasmas. 2010 (pp. 395–442). Springer, Berlin, Heidelberg.

[7] Shukla PK. Dusty Plasmas: Physics, Chemistry and Technological Impacts in Plasma Processing. In: Bouchoule A. Wiley, New York. 2000

[8] Bellan PM. Fundamentals of Plasma Physics. 1st ed. UK: Cambridge University Press; 2008

[9] Chen FF. Introduction to Plasma Physics and Controlled Fusion. 2nd ed. New York: Springer-Verlag. 2006; p. 200

[10] Vladimir EF, Gregor EM. Complex and Dusty Plasmas: From Laboratory to Space. Bosa Roca, United States: Taylor & Francis Inc, CRC Press Inc.; 2010

[11] Varma RK, Shukla PK, Krishan V. Electrostatic oscillations in the presence of grain-charge perturbations in dusty plasmas. Physical Review E. 1993;**47**(5): 3612. DOI: 10.1103/PhysRevE.47.3612

[12] Cui C, Goree J. Fluctuations of the charge on a dust grain in a plasma. IEEE Transactions on Plasma Science. 1994; **22**(2):151–158. DOI: 10.1109/27.279018

[13] Sharma SC, Sugawa M. The effect of dust charge fluctuations on ion cyclotron wave instability in the presence of an ion beam in a plasma cylinder. Physics of Plasmas. 1999;**6**(2):444–448 (1999).

[14] Merlino RL. Current-driven dust ion-acoustic instability in a collisional dusty plasma. IEEE Transactions on Plasma Science. 1997;**25**(1):60–65. DOI: 10.1109/27.557486

[15] Tribeche M, Zerguini TH. Current-driven dust ion-acoustic instability in a collisional dusty plasma with charge fluctuations. Physics of Plasmas. 2001;**8**(2):394–398. DOI: 10.1063/1.1335586

[16] Rao NN, Shukla PK, Yu MY. Dust-acoustic waves in dusty plasmas. Planetary and Space Science. 1990;**38**(4):543–546. DOI: 10.1016/0032-0633(90)90147-I

[17] Shukla PK, Silin VP. Dust ion-acoustic wave. Physica Scripta. 1992;**45**: 508. DOI: 10.1088/0031-8949/45/5/015

[18] Barkan A, Merlino RL, D'angelo N. Laboratory observation of the dust-acoustic wave mode. Physics of Plasmas. 1995;**2**(10):3563–3565. DOI: 10.1063/1.871121

[19] Morfill GE, Thomas H. Plasma crystal. Journal of Vacuum Science & Technology, A: Vacuum, Surfaces, and

Films. 1996;**14**(2):490–495. DOI: 10.1116/1.580113

[20] Barkan A, D'angelo N, Merlino RL. Experiments on ion-acoustic waves in dusty plasmas. Planetary and Space Science. 1996;**44**(3):239–242. DOI: 10.1016/0032-0633(95)00109-3

[21] Ali S. Potential distribution around a test charge in a positive dust-electron plasma. Frontiers of Physics. 2016;**11**(3): 115201. DOI: 10.1007/s11467-015-0545-2

[22] Bukhari SS. Ali, Rafique M, Mendonca JT. Twisted electrostatic waves in a self-gravitating dusty plasma. Contributions to Plasma Physics. 2017; **57**:404–413. DOI: 10.1002/ctpp.201700063

[23] Mendonça JT, Guerreiro A, Ali S. Photon bubbles in a self-gravitating dust gas: Collective dust interactions. The Astrophysical Journal. 2019;**142**(6pp): 872. DOI: 10.3847/1538-4357/aafe7e

[24] Pandey BP, Vranjes J. Physics of the dusty Hall plasmas. Physics of Plasmas. 2006;**13**(12):122106. DOI: 10.1063/1.2402148

[25] Kaufman HR. Technology of closed drift thrusters. AIAA Journal. 2012;**23** (1):78–86. DOI: 10.2514/3.8874

[26] Goebel DM, Katz I. Fundamentals of Electric Propulsion: Ion and Hall Thrusters. New York: Wiley; 2008

[27] Jahn RG. Physics of Electric Propulsion. New York: McGraw-Hill; 1968.

[28] Singh S, Malik HK, Nishida Y. High frequency electromagnetic resistive instability in a Hall thruster under the effect of ionization. Physics of Plasmas 2013;**20**:102–109 (1–7).

[29] Singh S, Malik HK. Growth of low frequency electrostatic and electromagnetic instabilities in a Hall thruster. IEEE Transactions on Plasma Science. 2011;**39**:1910–1918

[30] Singh S, Malik HK. Resistive instabilities in a Hall thruster under the presence of collisions and thermal motion of electrons. The Open Plasma Physics Journal. 2011;**4**:16–23

[31] Malik HK, Singh S. Resistive instability in a Hall plasma discharge under ionization effect. Physics of Plasmas. 2013;**20**:052115(1–8)

[32] Singh S, Malik HK. Role of ionization and electron drift velocity profile to Rayleigh instability in a Hall thruster plasma: Cutoff frequency of oscillations. Journal of Applied Physics. 2012;**112**:013307 (1–7)

[33] Malik HK, Singh S. Conditions and growth rate of Rayleigh instability in a Hall thruster under the effect of ion temperature. Physical Review E. 2011; **83**:036406 (1–8)

[34] Kapulkin A, Kogan A and Guelman M. Noncontact emergency diagnostics of SPT in flight. Acta Astronautica. 2004;**55**:109–119

[35] Fernandez E, Scharfe MK, Thomas CA, Gascon N, and Cappelli MA. Growth of resistive instabilities in $\vec{E} \times \vec{B}$ plasma discharge simulations. Physics of Plasmas. 2008;**15**:012102(1-10)

[36] Alcock M W and Keen B E. Experimental observation of the drift dissipative instability in an afterglow plasma. Physical Review A. 1971;**3**:1087–1096

[37] A. Kapulkin and M. M. Guelman. Low-frequency instability in near-anode region of Hall thruster. IEEE Transactions on Plasma Science. 2008; **36**:2082–2087

Chapter 6

Oseen's Flow Past Axially Symmetric Bodies in Magneto Hydrodynamics

Deepak Kumar Srivastava

Abstract

In the present technical note, drag on axially symmetric body for conducting fluid in the presence of a uniform magnetic field is considered under the no-slip condition along with the matching condition($\rho^2U^2 = H_0{}^2\mu^3\sigma$) involving Hartmans number and Reynolds number to define this drag as Oseen's resistance or Oseen's correction to Stokes drag is presented. Oseen's resistance on sphere, spheroid, flat circular disk (broadside) are found as an application under the specified condition. These expressions of Oseen's drag are seems to be new in magneto-hydrodynamics. Author claims that by this idea, the results of Oseen's drag on axially symmetric bodies in low Reynolds number hydrodynamics can be utilized for finding the Oseen's drag in magneto hydrodynamics just by replacing Reynolds number by Hartmann number under the proposed condition.

Keywords: stokes drag, Oseen's resistance, conducting fluid, magnetic field, Hartman number, Reynolds number

1. Introduction

There are many fluids like plasmas, liquid metals, salt water, and electrolytes etc. lies under the class of magneto hydrodynamics and attracted the attention of mechanical engineers, scientists and chemists for a longer period of time. The main significant quantity of magneto hydrodynamic fluid past an axially symmetric particle or object is the drag experienced by the stationary body or moving through the fluid.

It was **George Gabriel** Stokes [1] who gave the idea of Stokes drag on sphere by solving the Navier–Stokes equation combining with continuity equation under no-slip boundary condition by neglecting the convective inertia terms in the vicinity of spherical body. The then, this idea is known as Stokes law. This Stokes law or Stokes approximation is valid only in the vicinity of the body which breaks down at distance far away from the body. This breaks down of Stokes solution at far distance from the body is known as **Whitehead's paradox** [2]. It was **Oseen** [3], who pointed out the origin of Whitehead's paradox and suggest a scheme for its resolution (see [4]). In this scheme, Oseen has corrected the drag on the sphere, called Oseen's correction to Stokes drag, namely.

$$D = D_s\,[1 + (3/8)\,R], \tag{1}$$

where D_s is the classical Stokes drag and 'R' is the Reynolds number.

Chester [5] studied the effect of magnetic field on Stokes flow in a conducting fluid and modified the classical Stokes drag solution by magnetic field, which is uniform at infinity and is in the direction of flow of the fluid, given as

$$D = D_s\left\{1 + \frac{3}{8}M + \frac{7}{960}M^2 - \frac{43}{7680}M^3 + O\left(M^4\right)\right\}, \tag{2}$$

Where D_s is the classical Stokes drag and 'M' is the Hartmann number. He also proved that when the magnetic Reynolds number R_m, is small the magnetic field is essentially independent of the fluid motion. **Ludford** [6] discussed the effect of an aligned magnetic field on Oseen flow of a conducting fluid. **Payne and Pell** [7] have tackled the Stokes flow problem for a class of axially symmetric bodies and found the general expression of Stokes drag on axially symmetric bodies in terms of stream function. **Imai** [8] has discussed the flow of conducting fluid past bodies of various shapes. **Gotoh** [9] has discussed the magneto hydrodynamic flow past a sphere and calculated the drag on sphere. **Chang, I-Dee** [10] studied the problem of Stokes flow of a conducting fluid past an axially symmetric body in the presence of a uniform magnetic field and gave the formula of drag on axially symmetric body placed in the conducting fluid under the effect of uniform magnetic field. He utilizes the perturbation technique given by **Proudman and Pearson[11]**. In his Ph. D. thesis at Harvard University, **Blerkom** [12] studied the magneto-hydrodynamic flow of a viscous fluid past a sphere.

Brenner [13] calculated the Oseen resistance of a particle of arbitrary shape in terms of classical Stokes drag and Reynolds number 'R'. **Chester [14]** investigated the validity of the Oseen equations, for incompressible, viscous flow past a body, as an approximation to the Navier–Stokes equations. He determined the drag correctly to the first order in the Reynolds number, though the detailed velocity field is not correct to this order. Moreover, this force can be deduced simply from knowledge of the force on the body according to Stokes's approximation. He also analyzed the generalization of drag including the magneto-hydrodynamic effects when the fluid is conducting and the flow takes place in the presence of a magnetic field. **Kanwal** [15] has obtained the drag on solid bodies moving through the viscous and electrically conducting fluids. **Mathon and Ranger** [16] tackled the problem of magneto-hydrodynamic streaming flow past a sphere at low Hartmann numbers. **Bansal and Kumari** [17] have studied the MHD slow motion past a sphere and calculated the drag on sphere in both Stokes and Oseen's limits. **Datta and Srivastava** [18] proved a new form of Stokes drag on axially symmetric bodies based on geometric variables. **Venkatalaxmi et al.** [19] have obtained a general solution of Oseen equations based on the suggestions given by **Lamb** [20]. The Oseenlet is used for application purposes in their work. **Srivastava and srivastava** [21] calculated the Oseen's drag on axially symmetric bodies with the use of DS-conjecture given by **Datta and Srivastava** [18]. **Sellier and Aydin** [22] provided the fundamental free-space solutions for a steady axi-symmetric MHD viscous flow. **Ghosh et al.** [23] studied the effect of penetration of magnetic field on full magneto hydrodynamic flow past a circular cylinder. **Ibrahim and Tulu** [24] discussed the MHD boundary layer flow past a wedge with heat transfer and viscous effects of Nano fluid embedded in porous media. **Reza and Rajasekhar** [25] tackled the problem of shear flow over a rotating plate in the presence of magnetic field.

For in depth information regarding the classical Stokes drag and Oseen's drag on axi-symmetric bodies in relativistic fluid mechanics and magneto hydrodynamics, the books of Oseen [4] (in German Language), Happel and Brenner [26], Childress [27], Ferraro and Plumpton [28], Milne-Thompson [29], Cabannes [30], Mirela and Pop [31], Kim and Karrila [32] are referred by author.

2. Formulation of problem

We consider the equation of low Reynolds number flow of an incompressible conducting fluid past an axi-symmetric body in a magnetic field which is uniform at infinity. **Chester** [5] proved that when the magnetic Reynolds number R_m is small the magnetic field is essentially independent of fluid motion. For the case where the body and the fluid have nearly the same permeability, a uniform magnetic field will result i.e.$\mathbf{H}' = H_0\, \mathbf{i}$ = magnetic field at infinity. This indicates from the symmetry that there is no electric field, since for all such flows the electric currents form closed circuits. The governing equations and the no-slip boundary conditions for the present problem now becomes [10]

$$-\nabla p + \nabla^2 \mathbf{v} - M^2[\mathbf{v} - (\mathbf{v}.\mathbf{i})\mathbf{i}] = \mathbf{0}, \tag{3}$$

$$\nabla.\mathbf{v} = 0 \tag{4}$$

$$\mathbf{v} \to \mathbf{i} \text{ as } r \to \infty \left(r^2 = x^2 + y^2 + z^2\right), \tag{5}$$

$$\mathbf{v} = \mathbf{0} \text{ at the body.} \tag{6}$$

In Eqs. (3–5), all entities are non-dimensional and their abbreviations are as follows;

U = free-stream velocity,

a = characteristic length of body,

$$v = \frac{v'}{U}, p = \frac{a(p' - p'_\infty)}{\rho\nu U}, x = \frac{x'}{a}, \text{etc.},$$

$$Re = \frac{Ua}{\nu} = \text{Reynolds number},$$

$$R_m = Ua\mu\sigma = \text{magnetic Reynolds number},$$

$$M = \mu H_0 a \left(\frac{\sigma}{\rho\nu}\right)^{\frac{1}{2}} = \text{Hartmann number},$$

$$\mathbf{i} = \text{unit vector along x-direction.}$$

Other symbols have their usual meanings in electro-hydrodynamics and magneto-hydrodynamics. Primed entities are in physical units (as per [5, 10]).

Following the perturbation method given by **Proudman and Pearson** [11], Chang [10] has solved the above equations under the no-slip boundary conditions and obtained the drag on axially symmetric body in terms of classical Stokes drag D_s and Hartmann number as

$$D = D_s\left(1 + \frac{D_s}{16\pi\mu a U} M\right) + O(M^2), \tag{7}$$

where D_s is the Stokes drag for flow without magnetic field.

Now, in the section-4, we prove that the solution of drag given in Eq. (7) is Oseen's drag or Oseen's correction to Stokes drag by utilizing the idea of Oseen's resistance given by **Brenner** [13] for axially symmetric body under some specific constraints on dimensionless parameters responsible for the two flow configurations. The matching condition for Hartmann number and Reynolds number is provided as well.

3. Oseen's equations and Oseen's drag

Let us consider the axially symmetric arbitrary body of characteristic length 'a' placed along principal axis (x-axis, say) in a uniform stream U of viscous [4, 13] fluid of density ρ and kinematic viscosity ν. When particle Reynolds number Ua/ν is low, the steady motion of incompressible fluid around this axially symmetric body is governed by Stokes equations [26],

$$\mathbf{0} = -\left(\frac{1}{\rho}\right)\text{grad p} + \nu\,\nabla^2\mathbf{u}, \text{div}\,\mathbf{u} = 0, \tag{8}$$

subject to the no-slip boundary condition. It was **Oseen** in 1910, who pointed out the origin of Whitehead's paradox and suggest a scheme for its resolution (see [4]). In order to rectify the difficulty, Oseen proposed that uniformly valid solutions of the problem of steady streaming flow past a body at small particle Reynolds numbers could be obtained by solving the linear equations

$$(\mathbf{U}.\text{grad})\mathbf{u} = -\left(\frac{1}{\rho}\right)\text{grad p} + \nu\,\nabla^2\mathbf{u},\quad \text{div}\,\mathbf{u} = 0, \tag{9}$$

known as Oseen's equation. **Oseen [4]** obtained an approximated solution of his equations for flow past a sphere, from which he obtained the Stokes drag formula [**Happel and Brenner** [26] p. 44, (Eqs. 2-8)] under the no-slip conditions (Eqs. 5, 6) as

$$D = 6\,\pi\,\mu\,a\,U\left[1 + \frac{3}{8}\,\text{Re} + O\left(\text{Re}^2\right)\right], \tag{10}$$

where Re = ρUa/μ is bodies Reynolds number.

Based on Oseen's above idea and **Chang's[10]** expression of drag in terms of Hartmann number 'M', Brenner gave the expression of Oseen drag on axially symmetric body moving with equal velocity U and identical orientation through the unbounded fluid in terms of Reynolds number 'Re' as

$$D = D_s\left[1 + \frac{D_s}{16\pi\mu Ua}\,\text{Re}\right] + O\left(\text{Re}^2\right), \tag{11}$$

where 'a' is any characteristic particle or body dimension and Re = ρUa/μ is the particle Reynolds number.

4. Matching condition for Hartmann number and Reynolds number

In the expression of drag (Eq. 7) given by **Chang[10]**, the Hartmann number 'M' is treated as small. Similarly, in the expression of drag (Eq. 11) given by **Brenner** [13], the Reynolds number 'Re' is also treated as small. Now, we can define the drag D (Eq. 7) as Oseen's correction to classical Stokes drag D_s on axially symmetric body having characteristic length 'a' placed under uniform stream velocity U parallel to the principal axis(x-axis, say) when the two small dimensionless parameters M and Re matches to be equal i.e. M = Re provides

$$\frac{\rho U a}{\mu} = \mu H_0 a \left(\frac{\sigma}{\mu}\right)^{1/2},$$

$$\text{or } \rho^2 U^2 = H_0{}^2 \mu^3 \sigma. \tag{12}$$

Under this condition, drag on axially symmetric body in the presence of a uniform magnetic field described by **Chang[10]** is defined as Oseen's drag or Oseen's correction to Stokes drag in magneto-hydrodynamics. In the next section, we find the Oseen's drag on sphere and spheroid in terms of Hartmann number 'M' as an application which is the main task of interest for mechanical engineers.

5. Flow past sphere

We consider the sphere generated due to the revolution of circle of radius 'a' about axis of symmetry. The Oseen's drag on sphere of radius 'a' placed under conducting fluid of uniform velocity U and uniform magnetic field H_0 is given by (7) as

$$D = D_s\left(1 + \frac{D_s}{16\pi\mu a U} M\right) + O(M^2),$$

but for sphere, the classical Stokes drag $D_s = 6\pi\mu U a$, then, we have

$$D = D_s\left(1 + \frac{3}{8} M\right) + O(M^2), \tag{13}$$

which is in confirmation with Oseen's drag (Eq. 10) on sphere given by **Oseen** [4] and **Chester** [5] under the aforesaid condition(Eq. 10).

6. Flow past spheroid

6.1 Prolate spheroid

We consider the prolate spheroid generated by revolution of ellipse having semi-major axis length 'a' and semi-minor axis length 'b' about axis of symmetry. Stokes drag on prolate spheroid placed in uniform axial flow, with velocity U, parallel to axis of symmetry (x-axis) is given as (by utilizing DS conjecture given in [18])

$$D_s = \frac{16\,\pi\,\mu\,U\,a\,e^3}{\left[-2e + (1+e^2)\,\ln\frac{1+e}{1-e}\right]}. \tag{14}$$

Now, the Oseen's correction as well as the solution of Oseen's equation (Eq. 9) may be obtained for same prolate spheroid by substituting the value of Stokes drag (Eq. 14) in Brenner's formula (Eq. 11) under the matching condition (Eq. 10) as

$$\frac{D}{D_s} = 1 + \frac{16\,\pi\,\mu\,U\,a\,e^3}{16\,\pi\,\mu\,U\,a\left[-2e + (1+e^2)\ln\frac{1+e}{1-e}\right]}M + O(M^2),$$

$$= 1 + \frac{e^3}{\left[-2e + (1+e^2)\ln\frac{1+e}{1-e}\right]}M + O(M^2), \tag{15}$$

$$= 1 + \frac{3}{8}\left[1 - \frac{2}{5}e^2 - \frac{17}{175}e^4 \ldots\right]M + O(M^2), \tag{16}$$

where $M = \mu H_0 a\left(\frac{\sigma}{\rho\nu}\right)^{\frac{1}{2}}$ is Hartmann number and $R = \left(\frac{\rho U\,a}{\mu}\right)$ is the Reynolds number. The same solution may be re-written, when we take particle Reynolds number $R = \left(\frac{\rho Ub}{\mu}\right)$, by using $b/a = (1\text{-}e^2)^{1/2}$, as

$$= 1 + \frac{e^3}{\sqrt{1-e^2}\left[-2e + (1+e^2)\ln\frac{1+e}{1-e}\right]}M + O(M^2), \tag{17}$$

$$= 1 + \frac{3}{8}\left[1 + \frac{1}{10}e^2 + \frac{109}{1400}e^4 \ldots\right]M + O(M^2). \tag{18}$$

Equations (Eq. 16) and (Eq. 18) immediately reduces to the case of sphere (given in Eq. 13) in the limiting case as $e \to 0$. On the other hand, the closed form expressions (Eq. 15) and (Eq. 17) due to Oseen for prolate spheroid appears to be new for magneto hydrodynamics as no such type of expressions are available in the literature for comparison.

6.2 Oblate spheroid

We consider the oblate spheroid generated by revolution of ellipse having semi-major axis length 'b' and semi-minor axis length 'a' about axis of symmetry. Stokes drag on oblate spheroid placed in uniform axial flow, with velocity U, parallel to axis of symmetry (x-axis) is given as (by utilizing DS conjecture given in **[18]**)

$$D_s = \frac{8\,\pi\,\mu\,U\,a\,e^3}{\left[e\sqrt{1-e^2} - (1-2e^2)\sin^{-1}e\right]}. \tag{19}$$

Now, the Oseen's correction as well as the solution of Oseen's equation (Eq. 19) may be obtained for same oblate spheroid by substituting the value of Stokes drag (6.6) in Brenner's formula (Eq. 11) under the matching condition (Eq. 10) as

$$\frac{D}{D_s} = 1 + \frac{8\,\pi\,\mu\,U\,a\,e^3}{16\,\pi\,\mu\,U\,a\left[e\sqrt{1-e^2} - (1-2e^2)\sin^{-1}e\right]}M + O(M^2),$$

$$= 1 + \frac{e^3}{2\left[e\sqrt{1-e^2} - (1-2e^2)\sin^{-1}e\right]}M + O(M^2), \tag{20}$$

$$= 1 + \frac{3}{8}\left[1 - \frac{1}{10}e^2 - \frac{31}{1400}e^4 \ldots\right]M + O(M^2), \tag{21}$$

where $M = \mu H_0 a \left(\frac{\sigma}{\rho\nu}\right)^{\frac{1}{2}}$ is Hartmann number and $R = \left(\frac{\rho U\, a}{\mu}\right)$ is the Reynolds number. The same solution may be re-written, when we take particle Reynolds number $R = \left(\frac{\rho U\, b}{\mu}\right)$, by using $b/a = (1\text{-}e^2)^{1/2}$, as

$$= 1 + \frac{e^3}{2\sqrt{1-e^2}\left[e\sqrt{1-e^2} - (1-2e^2)\sin^{-1}e\right]} M + O(M^2), \tag{22}$$

$$= 1 + \frac{3}{8}\left[1 + \frac{2}{5}e^2 + \frac{61}{200}e^4 \ldots\right] M + O(M^2). \tag{23}$$

Equations (Eq. 21) and (Eq. 23) immediately reduces to the case of sphere (given in Eq. 13) in the limiting case as $e \to 0$. On the other hand, the closed form expressions (Eq. 20) and (Eq. 22) due to Oseen for oblate spheroid appears to be new as no such type of expressions are available in the literature for comparison.

7. Flat circular disk (broadside on)

Lamb [20] provided the Stokes drag on flat circular disk of radius 'a' placed broadside on facing towards the uniform stream of velocity U as.

$$D_s = 16\mu Ua. \tag{24}$$

Now, under the matching conditions (Eq. 10), the Oseen's drag on circular disk placed under the effect of magnetic field is given by Chang's rule (Eq. 7) in terms of Hartmann number as

$$D = D_s\left(1 + \frac{D_s}{16\pi\mu aU} M\right) + O(M^?)$$

or

$$D = 16\mu aU\left(1 + \frac{16\mu aU}{16\pi\mu aU} M\right) + O(M^2)$$

$$= 16\mu aU\left(1 + \frac{M}{\pi}\right) + O(M^2), \tag{25}$$

where $M = \mu H_0 a \left(\frac{\sigma}{\rho\nu}\right)^{\frac{1}{2}}$ is Hartmann number and $R = \left(\frac{\rho U\, a}{\mu}\right)$ is the Reynolds number. This drag immediately reduces to the classical one as $D = D_s$ defined by **Lamb [20]**. This Oseen's drag (Eq. 25) may also be reduced directly from oblate result (Eq. 20) by taking $e \to 1$ or $b \to 0$.

8. Conclusion

The problem of Oseen flow of an incompressible conducting fluid past axially symmetric body in the presence of a uniform magnetic field is tackled. The matching conditions are obtained by equating the small dimensionless Hartmann number and Reynolds number ensuring the **Chang's[10]** solution of drag to be Oseen's drag on same body in terms of Hartmann number and classical Stokes drag. Under no-slip boundary conditions, the closed form expressions are calculated to obtained the Oseen's drag on spheroid (prolate and oblate) and flat circular disk in terms of classical Stokes drag and Hartmann number. These expressions are further extended to the form containing powers of eccentricity 'e'. All forms reduces into the classical Oseen's drag on sphere of radius 'a' given **by Oseen's** [4] and **Chester [5, 14]**. These expressions of Oseen's drag are seems to be new in magneto-hydrodynamics. Following the same idea, the Oseen's drag may be calculated in terms of Hartmann number for other body configurations like deformed sphere, cycloidal body of revolution, egg-shaped body, cassini oval, hypocycloidal body etc.

Acknowledgements

Author conveys his sincere thanks to the authorities of B.S.N.V. Post Graduate College, Lucknow, UP, India, to provide basic infrastructure facilities and moral support throughout the preparation of this paper.

Author details

Deepak Kumar Srivastava
Department of Mathematics, B.S.N.V. Post Graduate College (K.K.V.), (University of Lucknow, Lucknow), Lucknow, U.P., India

*Address all correspondence to: dksflow@hotmail.com; srivastavadk1971@gmail.com

References

[1] G. G. Stokes (1851) On the effect of the internal friction of fluids on pendulums, Trans. Camb. Philos. Soc., vol. 9, p 8.

[2] Whitehead, A.N. (1889) Second approximations to viscous fluid motion, Quart. J. Math., vol. 23, no. 1, pp. 143–152.

[3] Oseen, C. W. (1910) Wber die Stokessche Formel und tiber eine Verwandte Anfgabe in der Hydrodynamik, Arkiv. Mat. Astron. Fys. vol. 6.

[4] Oseen, C. W. (1927) Neuere Methoden und Ergebnisse in der Hydrodynamik. Leipzig: Akademische Verlagsgesellschaft.

[5] Chester, W. E. (1957) The effect of magnetic field on Stokes flow in a conducting fluid, Journal of Fluid Mechanics, vol. 3, no. 3, pp. 304–308.

[6] Ludford, G. S. S. (1959) The effect of an aligned magnetic field on Oseen flow of a conducting fluid, Arch. Rational Mech. Anal., vol. 4, pp. 405–411.

[7] Payne, L.E. and Pell, W.H. (1960), The Stokes flow problem for a class of axially symmetric bodies, J. Fluid Mech. 7(4), pp. 529–549.

[8] Imai, I. (1960) On flows of conducting fluid past bodies, Rev. Mod. Phys., vol. 32, pp. 992–999.

[9] Gotoh, Kanefusa (1960) Magnetohydro dynamic flow past a Sphere, Journal of the Physical Society of Japan, vol. 15, no. 1, pp. 189–196.

[10] Chang, I-Dee (1960) Stokes flow of a conducting fluid past an axially symmetric body in the presence of a uniform magnetic field, vol. 9, no. 3, pp. 473–447.

[11] Proudman, I. and Pearson, J. R. A. (1957) Expansions at small Reynolds numbers for flow past a sphere and a circular cylinder. J. Fluid Mech., vol. 2, pp. 237–262.

[12] Blerkom, R. Von (1960) Magneto-hydrodynamic flow of a viscous fluid past a sphere, Ph. D. Thesis, Harvard University (1959), J. Fluid Mech. vol. 8, pp. 432–441.

[13] Brenner, Howard (1961) The Oseen resistance of a particle of arbitrary shape, J. Fluid Mech., vol. 11, pp. 604–610.

[14] Chester, W. (1962) On Oseen's approximation, J. Fluid Mech., vol. 13, pp. 557–569.

[15] Kanwal, R. P. (1969) Motion of solids in viscous and electrically conducting fluids, Journal of Mathematics and Mechanics, vol. 19, no. 6, pp. 489–513.

[16] Mathon, R. and Ranger, K. B. (1973) Magneto-hydrodynamic streaming flow past a sphere at low Hartmann numbers, Phys. Fluids, 16, pp. 485–490.

[17] Bansal, J. L. and Rama Kumari (1983) MHD slow motion past a sphere, Proc. Indian Acad. Sci. (Math. Sci.), vol. 92, pp. 157–166.

[18] Datta, S. and Srivastava, Deepak Kumar (1999) Stokes drag on axially symmetric bodies: a new approach, Vol. 109, Proceedings Math. Sci., Ind. Acad. Sci., No. 4, pp. 441–452.

[19] Venkatalaxmi, A., Padmavati, B. Sri and Amaranath, T. (2007) A general solution of Oseen equations, Fluid Dynamics Research, vol. 39, no. 7, pp. 595–606.

[20] H. Lamb (1945) Hydrodynamics, 6th ed., New York: Dover.

[21] Srivastava, Deepak Kumar and Srivastava, Nirmal (2013) Oseen's drag

on axially symmetric arbitrary particle: a new approach, Proceedings of the National Conference on "Role of Mathematics in Advancement of Science and Technology"[NCRMAST-2013] held at Deptt. of Mathematics, B. S.N.V. Post Graduate College, Lucknow, U.P., India, during Oct. 18–20, 2013, pp. 154–177.[ISBN No.-978–81–928613-0-2]

[22] Sellier, A. and Aydin, S. H. (2016) Fundamental free-space solutions of a steady axi-symmetric MHD viscous flow, European Journal of Computational Mechanics, vol. 27, no. 5–6, pp. 443–468.

[23] Ghosh, Somit, Sarkar, Subharthi, Sivakumar, R. and Sekhar, T.V.S. (2018) Full magneto hydro dynamic flow past a circular cylinder considering the penetration of magnetic field, Physics of Fluids, vol. 30, no. 8, 087102.

[24] Ibrahim, Wubshet and Tulu, Ayele (2019) MHD boundary layer flowpast a wedge with heat transfer and viscous effects of Nano fluid embedded in porous media, Mathematical Problems in Engineering (Hindawi) , vol. 2019, article id. 4507852.

[25] Reza, Motahar and Rajasekhar, G. P. (2020) Analysis and control of shear flow over a rotating plate in the presence of magnetic field, ZAMM, vol. 100, no. 4, pp. 1–15.

[26] Happel, J. and Brenner, H. (1964) Low Reynolds Number Hydrodynamics, Nijhoff, Dordrecht, The Nederlands.

[27] Childress, S. (1963) In fundamental topics in relativistic fluid mechanics and magneto hydrodynamics, Academic Press Inc., New York, USA.

[28] Ferraro, V. C. A. and Plumpton, L. (1966) An Introduction to Magneto-Fluid Mechanics, 2nd ed., Oxford University Press, Oxford, pp. 37–48.

[29] Milne-Thompson, L. M. (1968) Theoretical Hydrodynamics, Fifth Edition, Dover, NY.

[30] Cabannes, H. (1970) Theoretical magneto-fluid-dynamics, Academic Press, New York.

[31] Kohr, Mirela and Pop, Ioan I. (2004) Viscous Incompressible Flow for Low Reynolds Numbers, W.I.T. Press.

[32] Kim, Sangtae, Karrila, Seppo J. (2005). Microhydrodynamics: Principles and Selected Applications, Courier Corporation.

Chapter 7

A Note on Heat Transport with Aspect of Magnetic Dipole and Higher Order Chemical Process for Steady Micropolar Fluid

Assad Ayub, Hafiz A. Wahab, Zulqurnain Sabir and Adnène Arbi

Abstract

Heat transfer through non-uniform heat source/sink is the most significant aspect in view of many physical problems. Heat sink/source with heat transfer help to change the energy distribution in fluids, which consequently disturbs the particle deposition rate like as nuclear reactors, semiconductors and electronic devices. Further, also, the vital role of heat transfer is to enhance the thermal conductivity of micro sized solid particles in fluid. This study scrutinizes the heat transport of steady micropolar fluid via non-uniform heat sink/ source and mass transfer is scrutinized through higher order chemical reaction over a stretching surface with variable heat flux. Moreover, the velocity of micropolar fluid is studied by considering aspects of magnetic dipole and Newtonian heating; velocity slip conditions are also examined. The numerical results have been performed by using the well-known numerical shooting technique and comparison is performed with the Matlab built-in solver bvp4c. Geometrically explanation reveals the properties of numerous parameters that are the system parts. The observed outcomes show that the local skin-friction coefficient and Sherwood number values goes up with the increase of chemical reaction rate parameters and Schmidt numbers. Chemical reaction based parameters boosts up the rate of heat as well as mass transfer. The stress of wall couple increased by increasing the Schmidt and chemical parameters. Moreover, the plots of dimensionless parameters have been drawn, as well as some parameter results are tabulated.

Keywords: heat sink/source, heat transportation, magnetic dipole effect, Newtonian heating effect, micro polar fluid, slip velocity

1. Introduction

Magnetohydrodynamic (MHD) flow possesses real world applications for example, in the extrusion of a polymer sheet procedure, several product properties and significant control of cooling rate [1–3] and it controls under investigated physical state of the problem. Due to these important features of MHD flow, many researchers put their active attention towards the study of MHD and derived several numerical and theoretical results in mechanism of fluid flow. Andersson [4] did investigated about MHD flow of viscoelastic fluid with geometry of stretching surface. Thermal radiation and MHD flow are interrelated and have massive applications in industries to know impact of thermal radiation on MHD flow explored by Raptis et al. [5]. Khan et al. [6] examined the MHD flow of boundary layer using the electric behavior of the fluid due to the cause of stretching in an elastic plane surface along the magnetic field. Abergel et al. [7] did work on different problems related to different physical situations and existence of such solutions to control problems, like boundary and boundary control in a channel. Moreover, he provided basic numerical algorithms named as conjugate gradient and steepest descent methods. Agrawal et.al [8] did research to investigate the transfer of heat with MHD flow by applying an unvarying suction over stretching surface. These investigations were classified to the Newtonian fluids. However, a novel phase to evaluate the theory of fluid using MHD flow is categorized by Aliakbar et al. [9]. Chemical reactive flow and its relationship with magnetic dipole impact on Cross model is investigated by khan et al. [10]. Mahanthesh et al. [11] explore preparation of numerical results related to MHD nanofluid flow with bidirectional linear stretching surface. Babu et al. [12] published his study about MHD slip flow of nanofluid with mass transfer via thermophoresis and Brownian motion. Nadeem et al. [13] examined time dependent MHD three- dimensional flow due to stretching/shrinking sheet.

Class of fluid that exhibits microscopic effects arising the phenomenon of micromotion of fluid particles is called micropolar fluid. This fluid consists of rigid macromolecule of individual motion that supports stress, body moment by spin inertia. These fluids contain micro-constituents that are capable of undergoing the rotation and have several practical applications in different areas, like depicting the attitude of exotic lubricants, turbulent shear flow, colloidal suspensions of nanofluid flow, human and animal blood, exotic lubricants, additive suspensions, colloidal fluids, liquid crystal, real fluids with interruptions and so forth [14]. Soundalgekar et al. [15] described an outstanding analysis of the micropolar fluids (MPFs) along with its applications. In another work, Soundalgekar et al. [16] explored the suction/injection effects in the flow passing over a semi-infinite porous plate using (MHD) MPF flow. Hady et al. [17] obtained the analytical results for the heat transfer model to a MPF using a non-stretching sheet. Ishak et al. [18] worked on the heat transfer over a stretching surface together with variable or uniform surface in (MHD) MPFs. Hassanien et al. [19] numerically discussed the suction/ blowing effects on the heat/flow transfer using the MPF on a stretching surface. Hayat et al. [20] studied the two-dimensional mixed convection steady and stagnation point flow using (MHD) MPF on a stretching surface. Sajid et al. [21, 22] studied the true results for thin-film flows using MPF.

Vital role of thermal transport in various engineering problems, like as nuclear reactor cooling, metallurgical processes and continuous strips in which the performance of machine strongly dependent on the heat transfer rate and many hydrodynamic methods. Heat transfer through heat source/sink is most noteworthy aspect in view of many physical models. Heat generation/absorption can help to change the distribution energy in the fluid that consequently disturbs the particle deposition rate in the network like as semiconductors, nuclear reactors and electronic devices. Heat source/sink is assumed to be constant, temperature or space dependent. In this study, contains non-uniform heat sink/ source, i.e., temperature and space dependent heat source/sink. Motivated by the submissions of heat transfer with non-uniform heat source, numerous theoretical soundings have been discussed the heat transfer phenomenon in flows close to the stagnation point region [23]. Mabood et al. [24] used a shooting approach by considering the effects of thermal conductivity and variable viscosity using the MHD flow together with the transfer of heat in MPF through non uniform heat sink/source on a stretching sheet. The induction of flow is noticed because of an elastic sheet that is stretched back as well as forward. Reddy et al. [25] explored the heat transport via heat generation and mass transfer effects on MHD flow together with inclined porous plate. Ravidran et al. [26] implemented the non-uniform single/double effects of slot suction/injection into an unstable mixed convection of an electrically conducting. Ghadikolaei et al. [27] expressed detailed study with MHD flow heat transport inspired by the thermal radiations as well as heat generation over a porous stretching sheet. Sandeep et al. [28] scrutinized the non-uniform heat source/sink influences, chemical reaction and mass transfer on the mixed convection flow using a MPF along with viscous dissipation.

Most significance aspect of daily life is chemical reaction, and without chemical reaction there is no concept of daily life because chemical reactions appear in biomedical field, agriculture, photosynthesis, reproduction system, chemical industry, even earth is fertile with the chemical reactions and used in many engineering applications. Chemical reaction helps to transport the mass of fluid flow and many researchers did work on transport of mass with taking different models of fluid like Cross, Carreau, sisko and Maxwell model of non-Newtonian fluid. Numerical interpretation related mass transfer of 3D Cross fluid with chemical reaction is made by [29]. First order chemical reaction impact in MPF is studied by damesh et al. [30]. Das et al. [31] published their investigations on chemical reaction and thermal radiation on heat and mass transfer flow of MHD micropolar fluid with rotating frame of reference Further Magyari et al. [32] depicted that Combined effect of heat generation or absorption and first-order chemical reaction on micropolar fluid flows over a uniformly stretched permeable surface. Effects of higher order chemical reaction on micropolar fluid and Influence of thermophoresis and chemical reaction on MHD micropolar fluid flow is discussed by [33]. Sajid et al. investigate the effects of variable molecular diffusivity, nonlinear thermal radiation, convective boundary conditions, momentum slip, and variable molecular diffusivity on Prandtl fluid past a stretching sheet [34].

From many year scientists did a lot of work with different fluid models to investigate heat/ mass transport through heat generation/ absorption and activation energy respectively. But here in this manuscript we deal transport of heat of MPF via non uniform heat sink/ source because this transportation deal disposition rate of particle (space dependent) and mass transfer is carried out through chemical reaction. Further linear velocity of MPF is scrutinized by magnetic dipole aspect and also angular movement of said fluid is made in this struggle.

The remaining parts of the paper are organized as: Section 2 shows the problem formulation, Section 3is designed the methodology, Section 4 shows the results and discussion, while conclusion is drawn in the final Section.

2. Problem formulation

The problem is formulated by using the equations based on conservation of mass, momentum, angular momentum, energy equation and concentration equations, written as:

$$\nabla.V = 0, \tag{1}$$

$$\rho[\nabla.V]V = -\nabla.p + (\mu + \chi)\nabla^2 V + k\nabla \times \Omega + \rho f, \tag{2}$$

$$\rho j[\nabla.V]\Omega = (\alpha_0 + \beta_0 + \gamma_0)\nabla[\nabla\Omega] - \gamma_0\nabla \times [\nabla \times \Omega] + \chi\nabla \times V - 2k\Omega + \rho l, \tag{3}$$

$$\rho c_p[V.\nabla]T = k\nabla^2 T - \varphi + Q \tag{4}$$

$$[\nabla.j] = D\nabla^2 C. \tag{5}$$

Consider a mixed convective steady state incompressible MPF passing over a porous plate that shrinks and stretches towards the velocity $u = su_w(x)$, whereas, $u_w(x) = ax$, a is a dimensional constant towards the flow axis (**Figure 1**). The sheet is stretched with the speed that vary with distance ℓ, further consider the flow region is $y > 0$ and B_0 is the magnetic field towards the$y$-axis. Presence of boundary layer heat transport, non-uniform source of heat, viscous dissipation along with magnetic field is presented. The components of micro-rotation and velocity components are $(0, 0, w)$ and $(u, v, 0)$, respectively. The basic equations are accounted in the presence of heat source and viscous dissipation. The flow positions are x and y – axis along with the slip flow model, i.e., u-slip as well as the Newtonian heating are conditions as:

$$\frac{\partial u}{\partial x} + \frac{\partial v}{\partial y} = 0, \tag{6}$$

$$u\frac{\partial u}{\partial x} + v\frac{\partial u}{\partial y} = \left(\frac{\mu + \chi}{\rho}\right)\frac{\partial^2 u}{\partial y^2} + \frac{\chi}{\rho}\frac{\partial w}{\partial y} - \frac{\sigma\beta_0^2}{\rho}u + g_1\beta_C(C - C_\infty) + g_1\beta_T(T - T_\infty), \tag{7}$$

$$u\frac{\partial w}{\partial x} + v\frac{\partial w}{\partial y} = \left(\frac{\gamma}{\rho_j}\right)\frac{\partial^2 w}{\partial y^2} + \frac{\chi}{\rho_j}\left(2w + \frac{\partial w}{\partial y}\right), \tag{8}$$

$$u\frac{\partial T}{\partial x} + v\frac{\partial T}{\partial y} = \frac{k}{\rho C_p}\frac{\partial^2 T}{\partial y^2} + \frac{\mu}{\rho C_p}\left(\frac{\partial T}{\partial y}\right)^2 + \frac{q'''}{\rho C_p}, \tag{9}$$

$$u\frac{\partial C}{\partial x} + v\frac{\partial C}{\partial y} = D_m\frac{\partial^2 C}{\partial y^2} + D_m\frac{K_T}{T_m}\frac{\partial^2 T}{\partial y^2} - \xi(C - C_\infty). \tag{10}$$

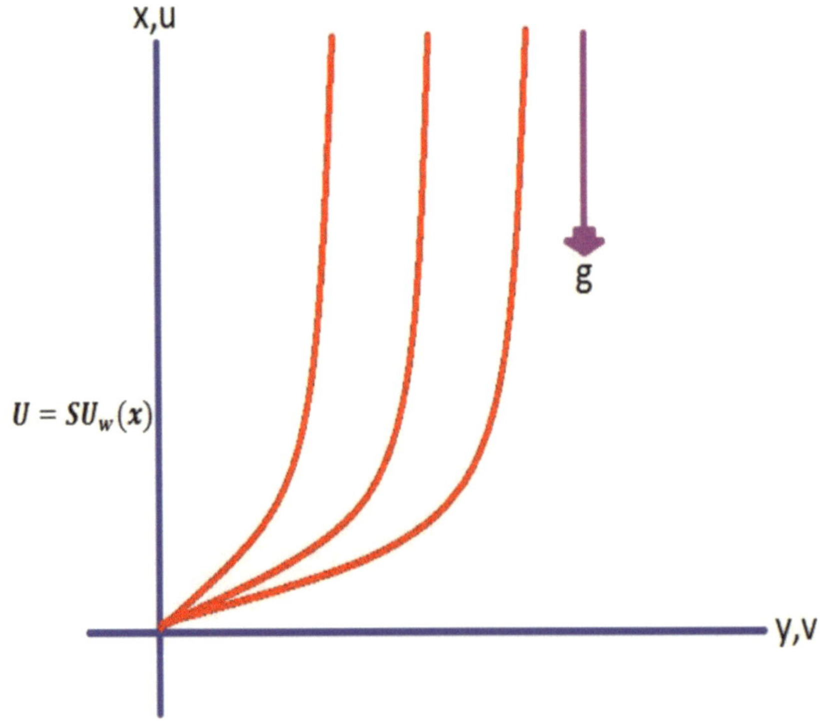

Figure 1.
Geometry of the problem.

To consider the spin gradient viscosity as

$$\gamma = \left(\mu + \frac{\chi}{2}\right)j = \mu\left(1 + \frac{K}{2}\right)j,\; K = \frac{\chi}{\mu} \text{ and } j = \frac{\nu}{a} \tag{11}$$

are the material parameter as well as micro inertia density, and $\nu = \frac{\mu}{\rho}$. Here q''' is the non-uniform heat source and defined as

$$q''' = (kU_w(x)/x\nu)\left[A^*(T_\infty)f'(\eta) + B^*(T - T_\infty)\right] \tag{12}$$

Moreover, $v_0 > 0$ and $v_0 < 0$ indicate the velocities based on suction as well as injection of the permeable plate.

The associated conditions are:

$$u = su_w(x) + u_{slip}, v = v_0, w = -n\frac{\partial u}{\partial y}, k\frac{\partial T}{\partial y} = -h_s T, C = C_w, \text{at } y = 0, \tag{13}$$

$$u \to 0, w \to 0, C \to C_\infty, T \to T_\infty, \text{ as } y \to \infty. \tag{14}$$

where n is constant and can be $0 \le n \le 1$, when, $n = 0$ leads to the concentration based micro-elements in MPF close to sheet that are do not rotate for $w = 0$, while $n = 1$ implies the turbulent flow. Moreover, u_{slip} represents the slip velocity and given as:

$$u_{slip} = \frac{2}{3}\left[\frac{3 - \varepsilon l^3}{\varepsilon} - \frac{3}{2}\frac{1 - l^2}{k_n}\right]d\frac{\partial u}{\partial y} - \frac{1}{4}\left[l^4 + \frac{2}{k_n^2}\left(1 - l^2\right)\right]d^2\frac{\partial^2 u}{\partial y^2}, \tag{15}$$

$$u_{slip} = A\frac{\partial u}{\partial y} + B\frac{\partial^2 u}{\partial y^2}, \tag{16}$$

where l is $\min\left[\left(\frac{1}{k_n}\right), 1\right]$ that goes to $0 < l \le 1$. The Knudsen number is k_n and ε represents the coefficient of momentum that lies in the range of $0 < \varepsilon \le 1$. The term d remains positive and shows the mean molecular free path, while B is accordingly negative. By presenting the following suitablc transformations as:

$$\Psi(x, y) = \sqrt{a\nu}xf(\eta), \phi(\eta) = \frac{C - C_\infty}{C_w - C_\infty}, \theta(\eta) = \frac{T - T_\infty}{T_\infty}, \eta = y\sqrt{\frac{a}{\nu}}, \tag{17}$$

$$w = axh(\eta)\sqrt{\frac{a}{\nu}}, u = \frac{\partial \Psi}{\partial y} = axf'(\eta), v = -\frac{\partial \Psi}{\partial x} = -\sqrt{a\nu}f(\eta), \tag{18}$$

where Ψ is called stream function. The momentum and heat equations are transformed to the governing momentum and heat transfer equations into the coupled ordinary differential equations as:

$$(1 + K)f''' + ff'' - f'^2 + Kh' - Mf' + \lambda\theta + \lambda A\phi = 0, \tag{19}$$

$$\left(1 + \frac{K}{2}\right)h'' + fh' - f'h - K(f'' + 2h) = 0, \tag{20}$$

$$\theta'' + \Pr f\theta' + Ec(f'')^2 + A^*f' + B^*\theta = 0, \tag{21}$$

$$\phi'' + Sc[f\theta' + Sr\theta'' + C_m\phi'] = 0, \tag{22}$$

The boundary conditions are

$$\begin{cases} f = f_w, f' = s + \alpha f'' + \beta f''', h = -nf'', \theta' = -\delta(1+\theta), \phi = 1 \text{ at } \eta = 0, \\ f'(\eta) \to 0, h(\eta) \to 0, \theta(\eta) \to 0, \phi(\eta) \to 0, \text{as } \eta = \infty. \end{cases} \tag{23}$$

Where the dimensionless parameters are defined as:
$K = \frac{\chi}{\mu}$, $\Pr = \frac{\mu C_p}{k}$, $Ec = \frac{a^2x^2}{c_p}$, $M = \frac{\sigma B_0^2}{\rho a}$, $\lambda = \frac{Gr_x}{Re_x}$, $Gr_x = \frac{g\beta_T T_\infty x}{a\nu}$, $C_m = \frac{\xi}{a}$, $\beta = \frac{h_s}{k}\sqrt{\frac{\upsilon}{a}}$,
$Sr = \frac{D_m k_T (T_w - T_\infty)}{T_m \nu (C_w - C_\infty)}$, $Du = \frac{D_m k_T \rho c_p (C_w - C_\infty)}{c_s c_p k (T_\infty)}$, $f_w = -[a\upsilon]^{-\frac{1}{2}}\nu_0$, $\mathrm{Re}_x = \frac{ax^2}{\upsilon}$, $\Lambda = \frac{g\beta_c (C_w - C_\infty)}{\beta_T T_\infty}$,
$a = A\left(\frac{a}{\upsilon}\right) > 0$ and $\beta = \left(\frac{Ba}{\upsilon}\right) < 0$.

The physical quantities based on skin-friction coefficient c_f is $c_{fx} = -\frac{\tau_w}{\rho U_w^2}$ the wall shear stress, τ_w is given as:

$$\tau_{wx} = \left| (\mu + \chi)\frac{\partial u}{\partial y} + \chi w \right|_{y=0}, \tag{24}$$

The value of c_f is given as:

$$c_{fx} \,\mathrm{Re}_x^{\,1/2} = -(1 + K - nK) f''(\eta)\big|_{y=0}, \tag{25}$$

Here R_{ex} is the Reynolds number. The skin-friction coefficient defined in Eq. (18) does not contain the micro rotation term. In the temperature field, the heat transfer rate is defined as:

$$Nu_x = -\frac{xq_w}{T - T_\infty}, \text{ where } q_w = \left(\frac{\partial T}{\partial T}\right)_{y=0}. \tag{26}$$

The local Nusselt number is shown as:

$$Nu_x(\mathrm{Re}_x)^{-0.5} = \delta\left(1 + \frac{1}{\theta(\eta)}\right)_{\eta=0}. \tag{27}$$

The couple stress is given as:

$$M_x = \frac{-m_w}{\rho x (ax)^2}, m_w = \left(\mu + \frac{\chi}{2}\right) j \left[\frac{\partial w}{\partial y}\right]_{y=0}, M_x \,\mathrm{Re}_x = -\left(1 + \frac{K}{2}\right) h'(0). \tag{28}$$

Furthermore, the mass diffusion flux and Sherwood number become as:

$$Sh_x = \frac{xS_m}{(C - C_w)}, \tag{29}$$

$$S_m = \left(\frac{\partial C}{\partial y}\right)_{y=0}. \tag{30}$$

Finally, Sherwood number becomes as:

$$Sh_x / (\mathrm{R}_{ex})^{-1/2} = (-\phi(\eta))_{\eta=0}. \tag{31}$$

The results of the above nonlinear equations have been performed by using a well-known shooting technique and comparison is performed with the bvp4c. The shooting technique is fast convergent scheme and have been used to solve the extensive applications of fluid dynamics [35–40]. For the implementations of the shooting scheme, the boundary value system has been converted into the initial value equations. The nonlinear Eqs. (19)-(22) take the form as:

$$\begin{cases} f = n_1,\ f' = n_2,\ f'' = n_3, \\ n_3' = \dfrac{1}{1+K}\left[Mn_2 - n_1n_3 + n_2^2 - Kn_5 - \lambda n_6 - \lambda\Lambda n_8\right], \\ n_4 = h,\ n_4' = n_5, \\ n_5' = \dfrac{1}{\left(1+\dfrac{K}{2}\right)}\left[n_2n_4 - n_1n_5 + K(2n_4 + n_3)\right], \\ n_6 = \theta, \theta' = n_7, \\ n_7' = \left[-\Pr n_1n_7 - Ecn_3^2 - (A^*n_2 + B^*n_6)\right], \\ n_8 = \phi, \phi' = n_9, \\ n_9' = \left[-Sc\left(n_1n_7 + Srn_7' + C_mn_8\right)\right]. \end{cases} \tag{32}$$

The concerned initial conditions are written as:

$$\begin{cases} n_1 = f_w, n_2 = s + \alpha n_3 + \beta n_3', n_4 = -nn_3, n_7 = -\delta(1+n_6), n_8 = 1 \text{ at } \eta = 0, \\ n_2 = n_4 = n_6 = n_8 = 0 \text{ as } \eta = \infty. \end{cases} \tag{33}$$

Matlab bvp4c technique procedure is given as

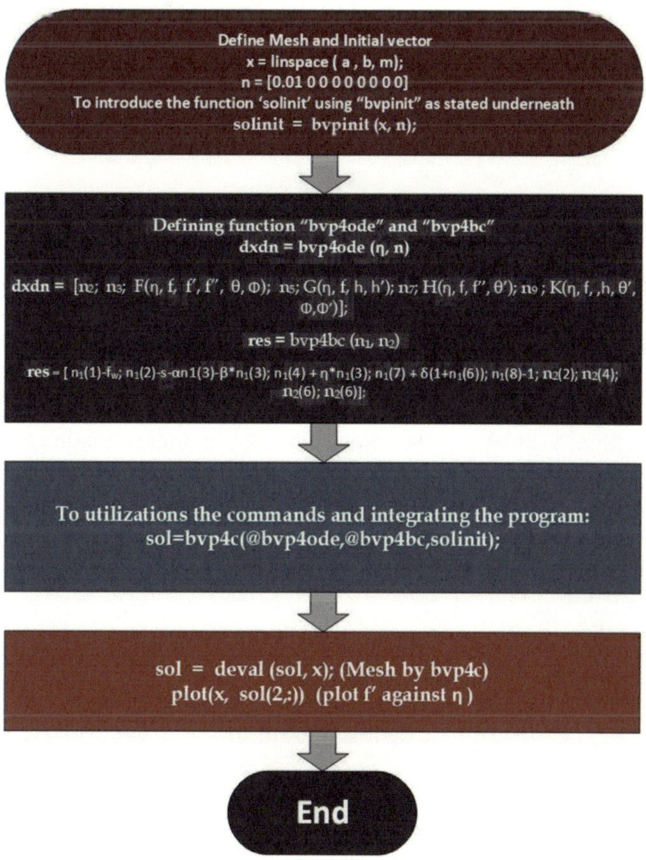

f_w	α	β	-$f''(0)$	-$f''(0)$	-$f''(0)$	-$f''(0)$
			Ref [27]	**Ref [28]**	**Bvp4c**	**Shooting**
2.0	0.5	−1	0.341213	0.3412	0.341214	0.341214
2.0	0.5	−2	0.203824	0.2038	0.203825	0.203825
2.0	1.0	−1	0.290548	0.2905	0.29057	0.29057
2.0	1.0	−2	0.184657	0.1846	0.18463	0.18463
3.0	0.5	−1	0.262681	0.2626	0.26281	0.26281
3.0	0.5	−2	0.147012	0.1470	0.14712	0.14712
3.0	1.0	−1	0.232017	0.2320	0.2314	0.2314
3.0	1.0	−2	0.136905	0.1369	0.13605	0.13605

Table 1.
Comparison value of $f''(0)$ for different values of the f_w.

For the satisfaction of the results, **Table 1** is provided is based on the literature results as well as the shooting and bvp4c for $f''(0)$ using numerous value of α by putting $K = 0$ and $M = 0$. The matching of the shooting and bvp4c results with the literature results [38, 39] depicts the satisfaction and validity of the scheme.

3. Results and discussions

In this section, the detail of the numerical results is presented to solve the system of nonlinear equations using shooting scheme. The effects of velocity profile, with physical parameters K, M, A, are examined, while Pr, Ec, A^*, B^*, are checked on temperature profile. Moreover, the influences of Sc, Sr, C_m is drawn on concentration profile through **Figures 2–14** as well as evaluation of physical quantities like c_f, Nu, M_x, Sh_x are provided.

3.1 Physical interpretation of parameters with velocity

The effects of parameter K, M, A, are examined by on the velocity profile and presented by **Figures 2–6**. Each parameter has its own impact for all the profiles along with its physical significance. As s is increasing, the sheet stretches due to this velocity increases $n \in (0, 1)$. When $n = 0$, then the MPF flow get closer to the sheet that are inept to rotate, likewise for $n = 1$ indicates the turbulent flow. The values of material constant 'K' enhance the velocity due to its materialistic properties. The velocity state decreases due to the Lorentz force by increasing the values of 'M'. Influence of the suction/injection parameter 'f_w' on the velocity component of the sheet shows an increment in the suction factor shows a decrease in the velocity together with the increment of the similarity values of the variable.

3.2 Physical interpretation of parameters with energy

The effects of different parameters are drawn on temperature profile are noticed in **Figures 7–12**. Temperature is decreasing for growing value of Pr, and B^* because of Pr reduces thermal conductivity and B^* is internal heat generation so negative values reduces the temperature. For also positive value of A^* temperature grows up. The energy intemperance exhibits a considerable increase with the wall temperature. This is reliable with the physical state because of elastic deformation

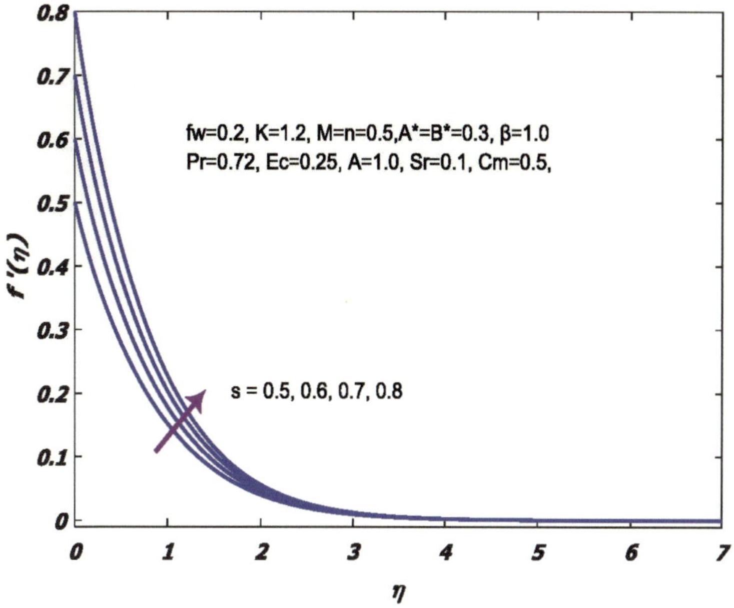

Figure 2.
s effects on f profile.

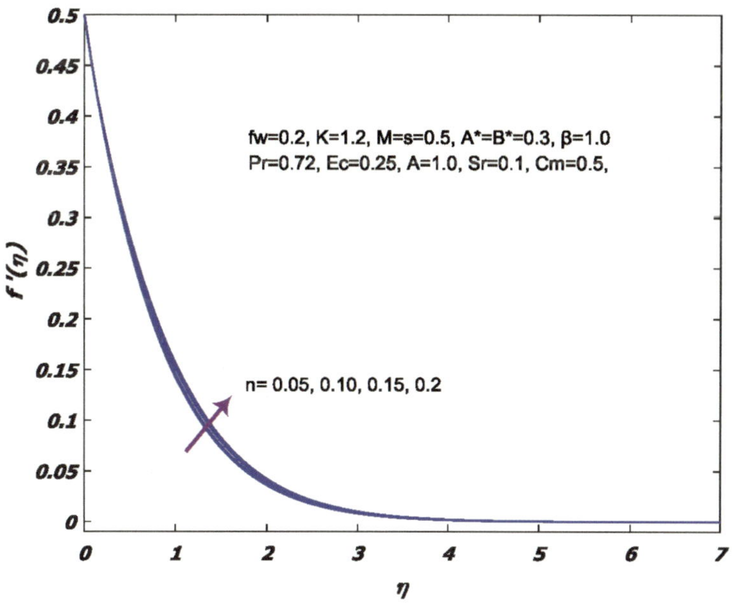

Figure 3.
n effects on f profile.

work, ohmic and frictional heating are considered that become the cause of incrementing the thermal based boundary layer.

3.3 Physical interpretation of parameters with concentration

Figures 11–14 depicts the angular velocity of MPF with attached parameters, for increasing value of Pr, n, M angular velocity increases. For growing Pr temperature

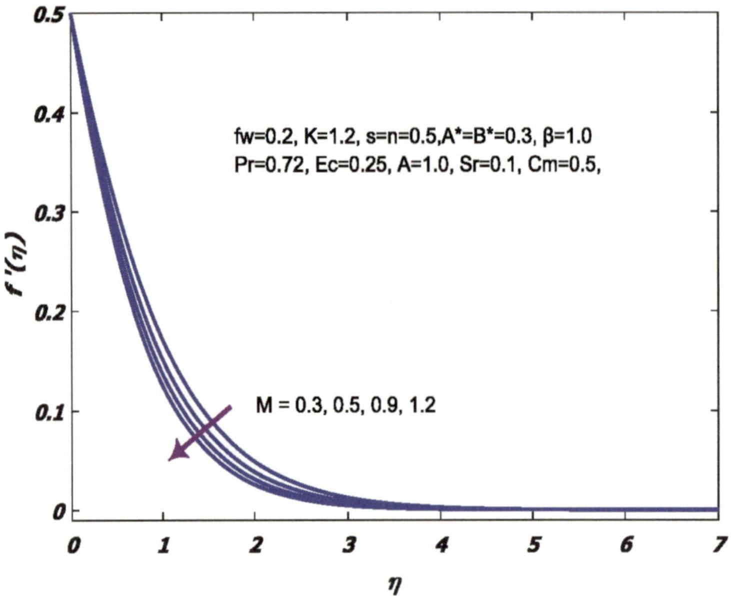

Figure 4.
M effects on f profile.

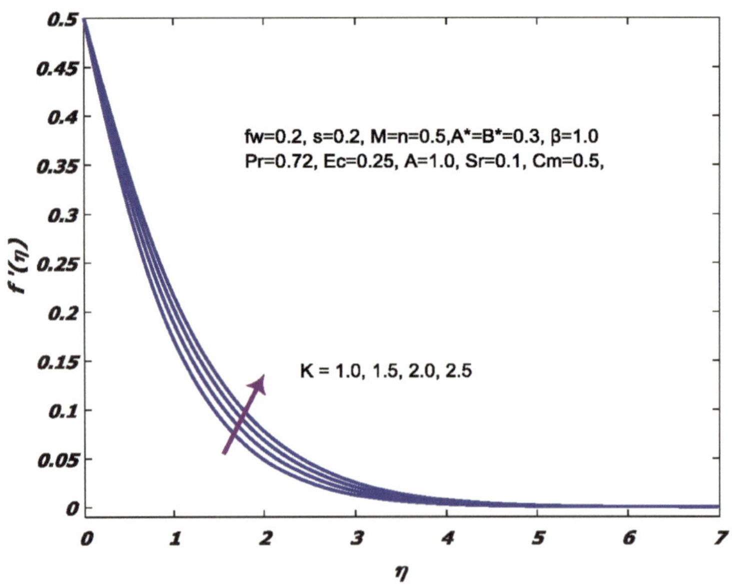

Figure 5.
K effects on f profile.

loses due to this movement of particle slows down due to this rotation gets down, and n is constant and there is no rotation when $n = 0$, so increase in n results growth in angular velocity. Similarly, a greater value of M produces Lorentz force due to this linear velocity downs but angular velocity uplifts. As the material parameter increases, it is observed that the boundary layer thickness increases due to this fact angular velocity decreases. The concentration profile became down with increment

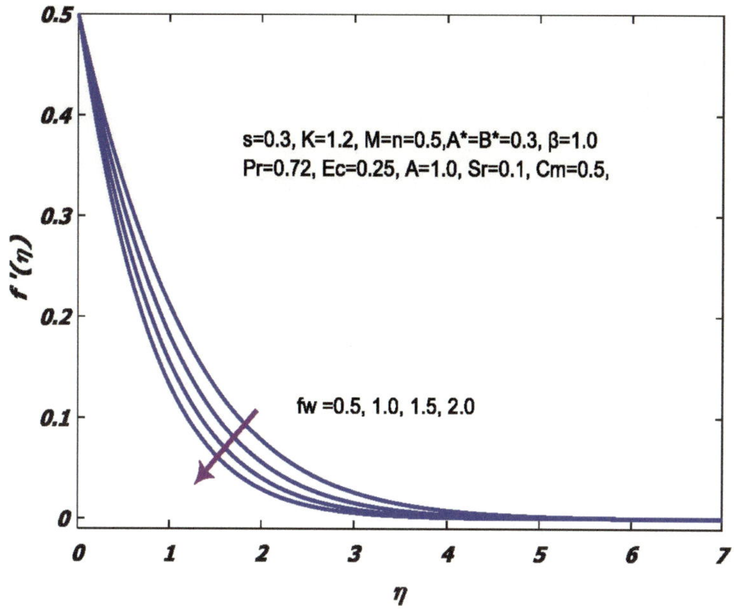

Figure 6.
f_w effects on f profile.

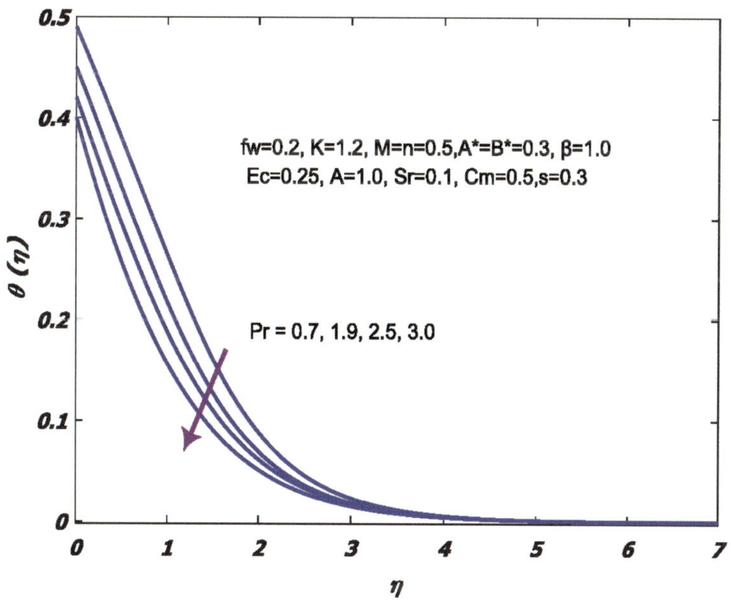

Figure 7.
Pr effects on θ profile.

of Pr, Sc, Sr, f_w and C_m. The increasing values of 'Pr' depicts that temperature decreases, as a result decrement is noticed in concentration. The concentration along with thickness of boundary layer decreases by enhancing the 'Sc'. The Soret term, i.e., 'Sr' shows the temperature gradients effects on the profile of concentration. it is noticed that increment in 'Sr', temperature together with concentration increases (**Figures 15–19**).

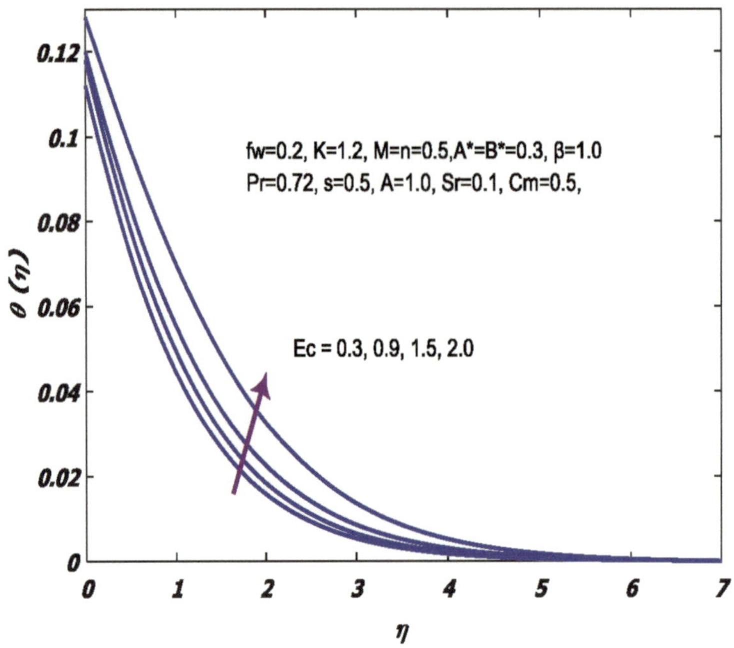

Figure 8.
E_c effects on θ profile.

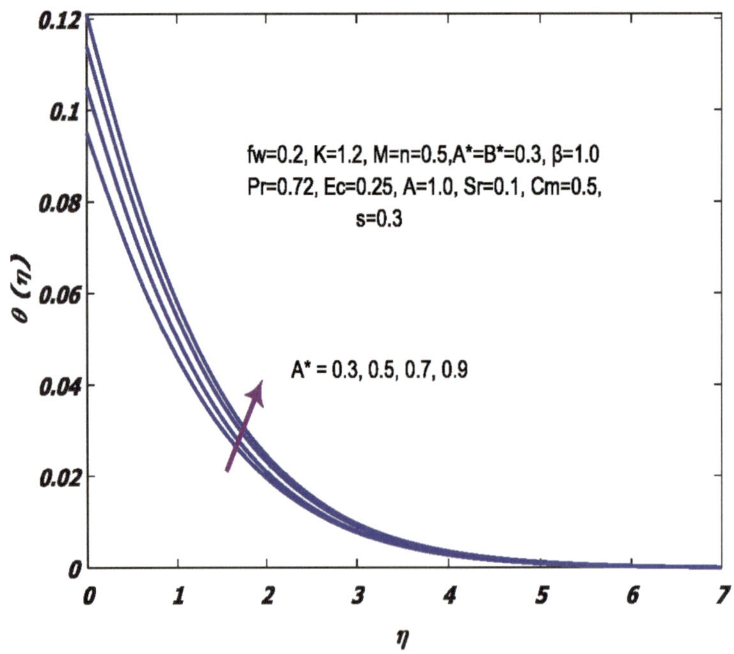

Figure 9.
A^* effects on θ profile.

3.4 Physical quantities interpretation

Skin friction is called the rate of heat transfer that shows the increment for greater value of K, Sc, C_m and decreases for growing values of A and f_w. Couple

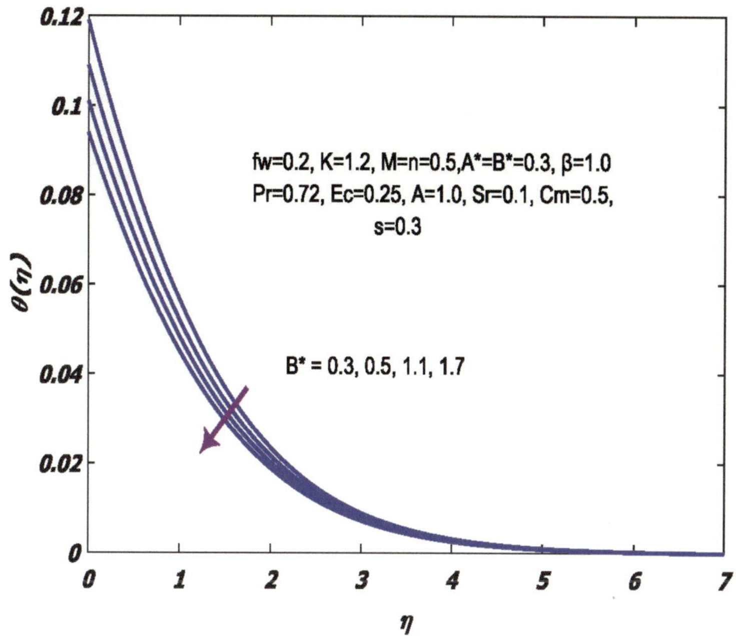

Figure 10.
B^* effects on θ profile.

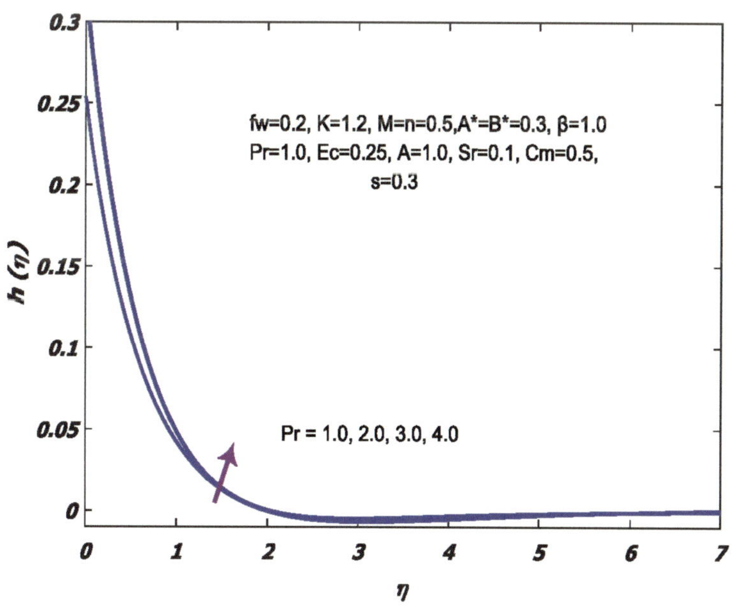

Figure 11.
Pr effects on h profile.

stress $M_x \mathrm{Re}_x$ values vary with parameter K. In this study, an increment is found in the couple stress $M_x \mathrm{Re}_x$ for growing value of K. Local mass diffusion flux namely Sh_x goes down for rising value of K and numerical result of the present study in **Table 2** is listed.

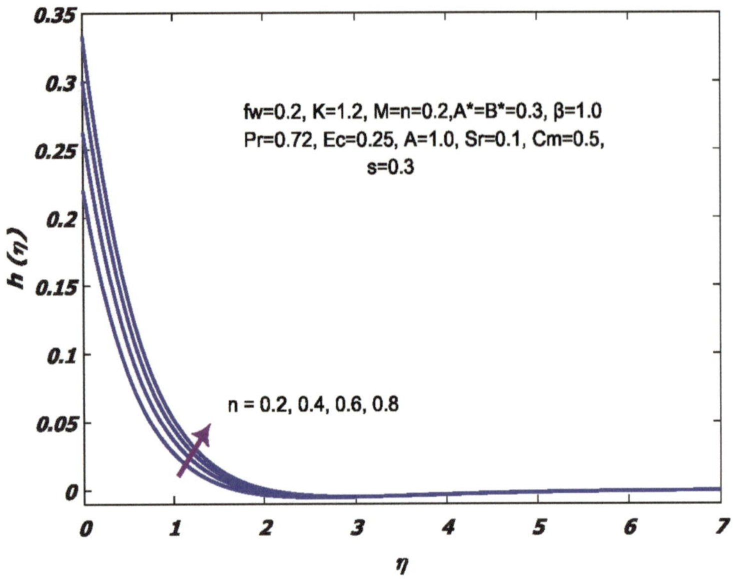

Figure 12.
n effects on h profile.

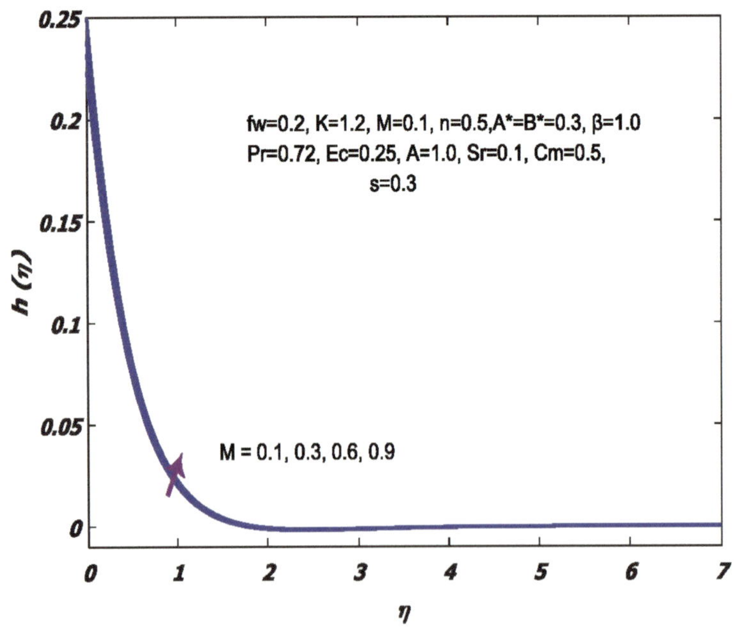

Figure 13.
M effects on h profile.

4. Concluding remarks

The key purpose of the current work is to discuss the effects of heat transportation and source/sink of heat with magnetic effect on boundary layer MPF. This study elaborates that angular velocity and linear velocity of MPF with different facts which impact on temperature and concentration of said fluid. For the

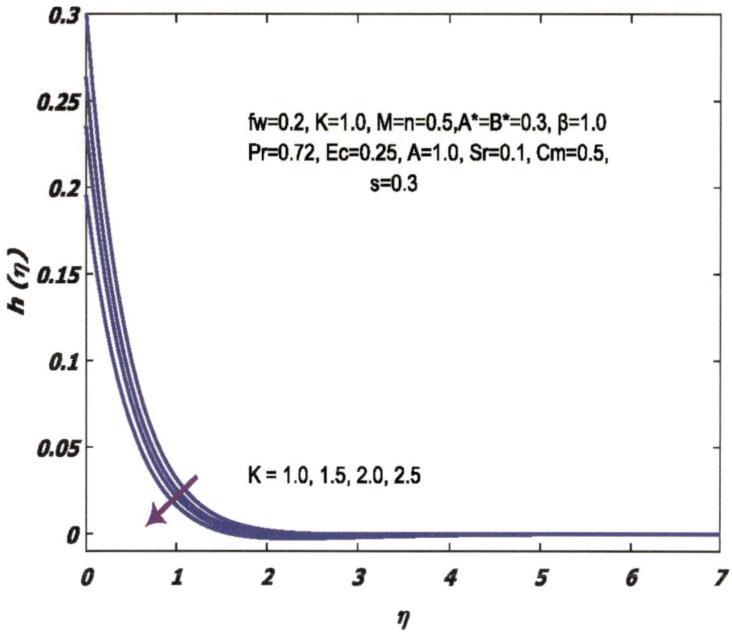

Figure 14.
K effects on h profile.

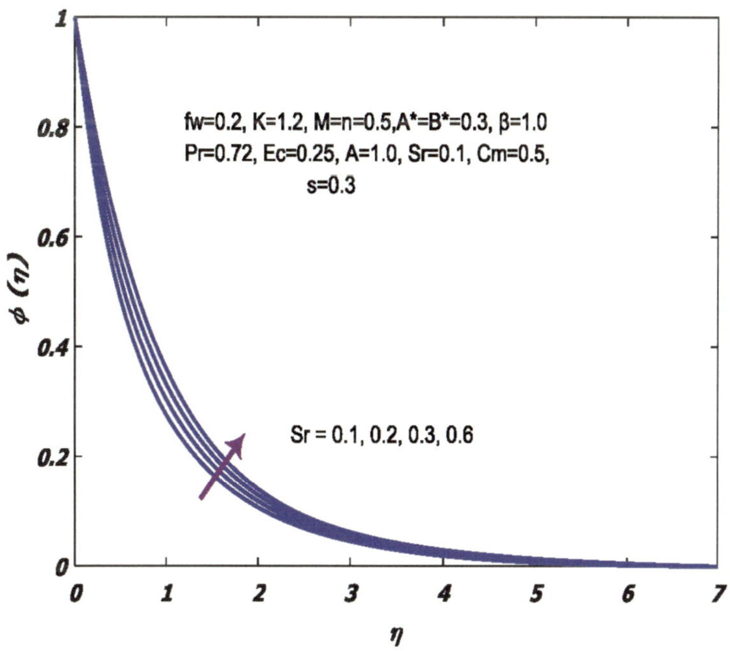

Figure 15.
S_r effects on ϕ profile.

numerical purpose, the shooting scheme has been implemented and comparison of the results with bvp4c is presented. Moreover, main key points of study are provided as:

1. Velocity of flow raises for growing of s, n, K.

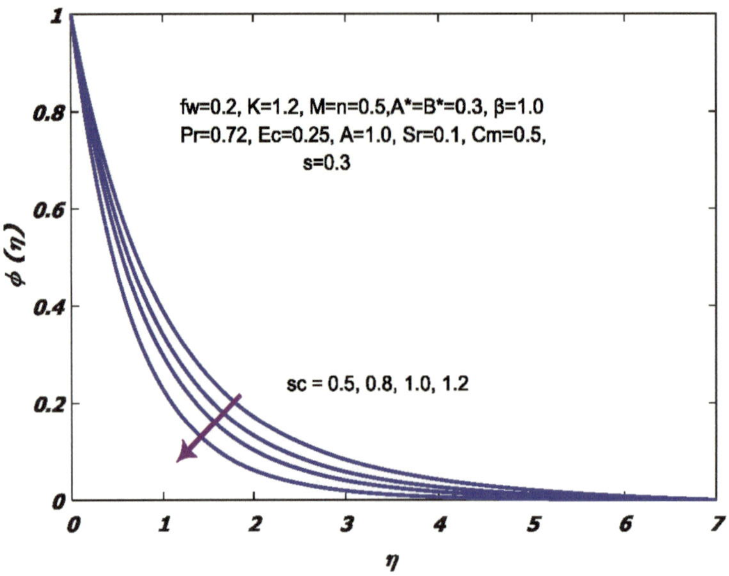

Figure 16.
Sc effects on φ profile.

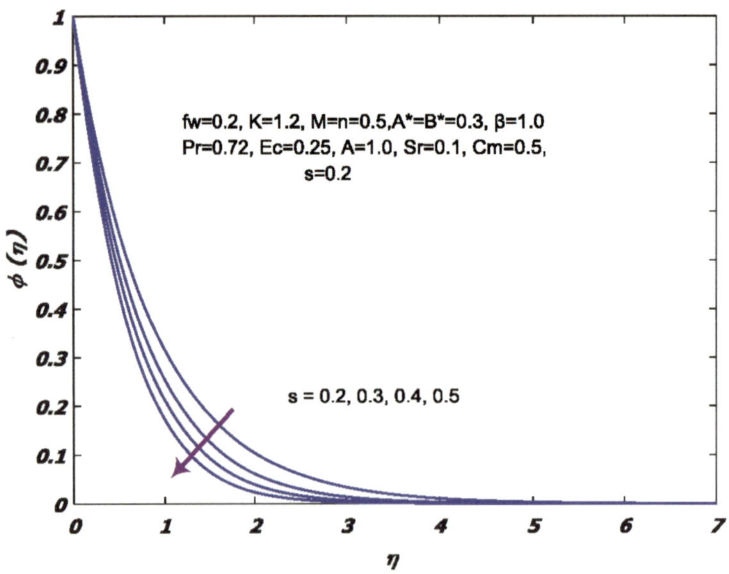

Figure 17.
s effects on φ profile.

2. Velocity of flow downs for growing of M, f_w.

3. For also positive value of A^* and increasing value of Ec temperature grows up

4. Mass field and corresponding boundary layers thickness downs by increasing the 'Sc'.

5. An increasing in 'Sr' causes a increase in the concentration and temperature through the boundary layer.

DOI: http://dx.doi.org/10.5772/intechopen.95302

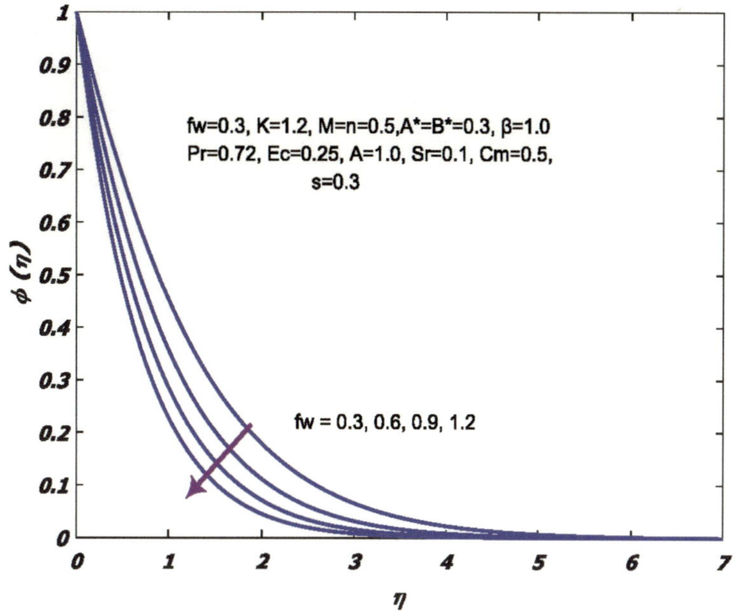

Figure 18.
f_w effects on ϕ profile.

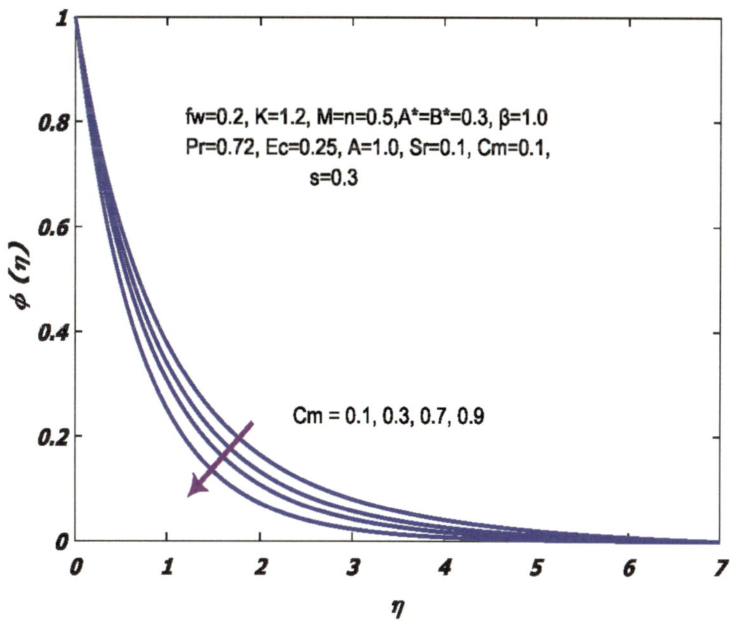

Figure 19.
Cm effects on ϕ profile.

K	A	Sc	C_m	f_w	$c_{fx}\,\mathrm{Re}_x^{1/2}$	$Nu_x(\mathrm{R_{ex}})^{-1/2}$	$M_x\,\mathrm{Re}_x$	$Sh_x/(\mathrm{R_{ex}})^{-1/2}$
0					0.3455454	1.658233	0.164237	0.7418682
0.5					0.4135856	1.668674	0.193939	0.7471964
1.0					0.482298	1.678695	0.2210438	0.7525225

K	A	Sc	C_m	f_w	$c_{fx}\,\mathrm{Re}_x^{\;1/2}$	$Nu_x(\mathrm{R_{ex}})^{-1/2}$	$M_x\,\mathrm{Re}_x$	$Sh_x/(\mathrm{R_{ex}})^{-1/2}$
1.5					0.5488359	1.678766	0.2460082	0.7575476
2.0					0.6130726	1.688377	0.2691283	0.7622267
0.5	0.5				0.4189335	1.668688	0.1939364	0.9283288
	1.0				0.41838978	1.688139	0.193138	0.9291135
	1.5				0.4178619	1.7081801	0.192349	0.9298976
	2.0				0.4173480	1.738553	0.191564	0.9306813
	0.5	0.5			0.4135858	1.668739	0.162426	0.7471966
		1.0			0.4319673	1.668286	0.169963	1.148792
		1.5			0.4431084	1.668264	0.174907	1.488861
		2.0			0.4507155	1.6687357	0.1785145	1.800042
		0.5	0.0		0.4014076	1.6683189	0.157579	0.1215778
			0.5		0.4135857	1.668763	0.162424	0.7471969
			1.0		0.4216537	1.6685654	0.1657039	1.148795
			1.5		0.4274917	1.668875	0.1681314	1.488867
			2.0	1.0	0.431996	1.748467	0.170059	1.800044
			0.5	0.5	0.420988	1.288242	0.212292	0.799846
				0.0	0.4189334	1.998349	0.193938	0.842233
				−0.5	0.4147433	1.808545	0.174625	0.885127
				−1.0	0.3915482	1.69878	0.1535629	0.9283398

Table 2.
Behavior of skin friction, Nusselt number, couple stress, and Sherwood number.

Nomenclature of parameters and short terms

f	Body force
l	Body couple
V	translational vector
Ω	micro-rotation vector
K	Thermal conductivity
p	Pressure
$\alpha_0, \beta_0, \gamma_0, \chi$, K	Material constants
c_s	Concentration susceptibility
M	Magnetic parameter
j	Micro-inertia
μ	dynamic viscosity
φ	dissipation function
τ	Ratio parameter
C	Concentration of fluid.
T	Temperature of the fluid
V	translational vector
Ω	micro-rotation vector
T_∞	Infinite temperature
A	Usual constant

A	Constant
a	first order slip flow parameter
T_w	Temperature of the plate
c_p	Specific heat
ρ	Fluid density
D_m	coefficient of mass diffusivity
α	Thermal diffusivity
Du	Dufour number,
β	Second order slip flow parameter
MPF	Micropolar fluid
$q'r'$	Non-uniform heat source
A^*, B^*	Coefficients of space and temperature
D	Mass diffusivity
K_T	Thermal diffusion ratio
D_m	Coefficient of mass diffusivity
T_m	Mean fluid temperature.
N	Constant
α^*	Slip coefficient
w	Micro rotation component
h	Heat transfer coefficient
Pr	Prandtl number
G	Micro rotation parameter
Ec,	Eckert number
σ	Reaction rate parameter
β_o	Strength of magnetic field.
Sc	Schmidt number
λ^A	Activation energy parameter
Bi_θ	Thermal Biot number
α	slip parameter
γ_1	Thermal concentration parameter.
Sr	Soret number
f_w	Suction or injection parameter
Sh_x	Sherwood number
ν	Kinematic viscosity
c_f	Skin-friction coefficient
R_{ex}	Local Reynold number
N_{ux}	Nusselt number
m_w	couple-stress
δ	Newtonian heating parameter
MHD	Magnetohydrodynamic

Author details

Assad Ayub[1], Hafiz A. Wahab[1], Zulqurnain Sabir[1] and Adnène Arbi[2,3]*

1 Department of Mathematics and Statistics, Hazara University, Manshera, Pakistan

2 Laboratory of Engineering Mathematics (LR01ES13), Tunisia Polytechnic School, University of Carthage, Tunisia

3 Department of Advanced Sciences and Technologies at National School of Advanced Sciences and Technologies of Borj Cedria, University of Carthage, Tunisia

*Address all correspondence to: adnen.arbi@gmail.com

References

[1] Herdrich, G., Auweter-Kurtz, M., Fertig, M., Nawaz, A., & Petkow, D. (2006). MHD flow control for plasma technology applications. *Vacuum*, *80* (11-12), 1167-1173.

[2] Smolentsev, S., Badia, S., Bhattacharyay, R., Bühler, L., Chen, L., Huang, Q., ... & Mistrangelo, C. (2015). An approach to verification and validation of MHD codes for fusion applications. *Fusion Engineering and Design*, *100*, 65-72.

[3] Sabir, Z., Ayub, A., Guirao, J. L., Bhatti, S., & Shah, S. Z. H. (2020). The Effects of Activation Energy and Thermophoretic Diffusion of Nanoparticles on Steady Micropolar Fluid along with Brownian Motion. *Advances in Materials Science and Engineering*, *2020*.

[4] Andersson, H. I. (1992). MHD flow of a viscoelastic fluid past a stretching surface. *Acta Mechanica*, *95*(1-4), 227-230.

[5] Raptis, A., Perdikis, C., & Takhar, H. S. (2004). Effect of thermal radiation on MHD flow. *Applied Mathematics and computation*, *153*(3), 645-649.

[6] Khan, M., Hussain, M., & Azam, M. (2016). Magnetohydrodynamic flow of Carreau fluid over a convectively heated surface in the presence of non-linear radiation. *Journal of magnetism and magnetic materials*, *412*, 63-68.

[7] Abergel, F., & Temam, R. (1990). On some control problems in fluid mechanics. *Theoretical and Computational Fluid Dynamics*, *1*(6), 303-325.

[8] Agrawal, P., Dadheech, P. K., Jat, R. N., Bohra, M., Nisar, K. S., & Khan, I. (2020). Lie similarity analysis of MHD flow past a stretching surface embedded in porous medium along with imposed heat source/sink and variable viscosity. *Journal of Materials Research and Technology*, *9*(5), 10045-10053.

[9] Aliakbar, V., Alizadeh-Pahlavan, A., & Sadeghy, K. (2009). The influence of thermal radiation on MHD flow of Maxwellian fluids above stretching sheets. *Communications in Nonlinear Science and Numerical Simulation*, *14*(3), 779-794.

[10] Khan, M. I., Hayat, T., Khan, M. I., & Alsaedi, A. (2018). Activation energy impact in nonlinear radiative stagnation point flow of Cross nanofluid. *International Communications in Heat and Mass Transfer*, *91*, 216-224.

[11] Mahanthesh, B., Gireesha, B. J., Gorla, R. R., Abbasi, F. M., & Shehzad, S. A. (2016). Numerical solutions for magnetohydrodynamic flow of nanofluid over a bidirectional non-linear stretching surface with prescribed surface heat flux boundary. *Journal of Magnetism and Magnetic Materials*, *417*, 189-196.

[12] Babu, M. J., & Sandeep, N. (2016). Three-dimensional MHD slip flow of nanofluids over a slendering stretching sheet with thermophoresis and Brownian motion effects. *Advanced Powder Technology*, *27*(5), 2039-2050.

[13] Nadeem, S., Haq, R. U., & Akbar, N. S. (2013). MHD three-dimensional boundary layer flow of Casson nanofluid past a linearly stretching sheet with convective boundary condition. *IEEE Transactions on Nanotechnology*, *13* (1), 109-115.

[14] T. Armin, M.A. Turk, N.D. Sylvester, Application of microcontinuum fluid mechanics, Int. J. Engng. Sci. 12 (1974) 273–279.

[15] V.M. Soundalgekar, H.S. Takhar, Flow of a micropolar fluid on a

continuous moving plate, Int. J. Engng. Sci. 21 (1983) 961– 965.

[16] F.M. Hady, Short communication on the solution of heat transfer to micropolar fluid from a non-isothermal stretching sheet with injection, Int J. Num. Meth. Heat Fluid Flow 6 (1996) 99–104.

[17] A. Ishak, R. Nazar, I. Pop, Heat transfer over a stretchingsurface with variable surface heat flux in micropolar fluids, Phys. Lett. A 372 (2008) 559–561.

[18] I.A. Hassanien, R.S.R. Gorla, Heat transfer to a micropolar fluid from a non-isothermal stretching sheet with suction and blowing, Acta Mech. 84 (1990) 191–199.

[19] T. Hayat, Z. Abbas, T. Javed, Mixed convection flow of a micropolar fluid over a non-linear stretching sheet, Phys. Lett. A372 (2008) 637–647.

[20] T. Hayat, T. Javed, Z. Abbas, MHD flow of a micropolar fluid near a stagnation-point towards a non-linear stretching surface, Nonlinear Anal.: Real World Appl. 10 (2009) 1514–1526.

[21] M. Sajid, N. Ali, T. Hayat, On exact solutions for thin film flows of a micropolar fluid, Commun. Nonlinear Sci. Num. Simul. 14 (2009) 451–461.

[22] M. Sajid, Z. Abbas, T. Hayat, Homotopy analysis for boundary layer flow of a micropolar fluid through a porous channel, Appl. Math. Model. 33 (2009) 4120–4125.

[23] Alsaedi, A., Awais, M., & Hayat, T. (2012). Effects of heat generation/ absorption on stagnation point flow of nanofluid over a surface with convective boundary conditions. *Communications in Nonlinear Science and Numerical Simulation*, *17*(11), 4210-4223.

[24] Mabood, F., Ibrahim, S. M., Rashidi, M. M., Shadloo, M. S., & Lorenzini, G. (2016). Non-uniform heat source/sink and Soret effects on MHD non-Darcian convective flow past a stretching sheet in a micropolar fluid with radiation. *International Journal of Heat and Mass Transfer*, *93*, 674-682.

[25] Reddy, M. G., & Reddy, N. B. (2011). Mass transfer and heat generation effects on MHD free convection flow past an inclined vertical surface in a porous medium.

[26] R. Ravindran, M. Ganapathirao, I. Pop, Effects of chemical reaction and heat generation/absorption on unsteady mixed convection MHD flow overa vertical cone with non-uniform slot mass transfer, Int.J.Heat & Mass Transfer 73 (2014) 743-751.

[27] Ghadikolaei, S. S., Hosseinzadeh, K., Ganji, D. D., & Jafari, B. (2018). Nonlinear thermal radiation effect on magneto Casson nanofluid flow with Joule heating effect over an inclined porous stretching sheet. *Case Studies in Thermal Engineering*, *12*, 176-187.

[28] N, Sandeep, C, Sulochana, Dual solutions for unsteady mixed convection flow of MHD micropolar fluid over a stretching /shrinking sheet with nonuniform heat source/sink, JESTECH. 18 (2015) 738-745.

[29] Hayat, T., Khan, M. I., Tamoor, M., Waqas, M., & Alsaedi, A. (2017). Numerical simulation of heat transfer in MHD stagnation point flow of Cross fluid model towards a stretched surface. *Results in physics*, *7*, 1824-1827.

[30] Damseh, R. A., Al-Odat, M. Q., Chamkha, A. J., & Shannak, B. A. (2009). Combined effect of heat generation or absorption and first-order chemical reaction on micropolar fluid flows over a uniformly stretched permeable surface. *International Journal of Thermal Sciences*, *48*(8), 1658-1663.

[31] Das, K. (2011). Effect of chemical reaction and thermal radiation on heat

and mass transfer flow of MHD micropolar fluid in a rotating frame of reference. *International journal of heat and mass transfer*, *54*(15-16), 3505-3513.

[32] Magyari, E., & Chamkha, A. J. (2010). Combined effect of heat generation or absorption and first-order chemical reaction on micropolar fluid flows over a uniformly stretched permeable surface: the full analytical solution. *International Journal of Thermal Sciences*, *49*(9), 1821-1828.

[33] Yokuş, A., & Gülbahar, S. (2019). Numerical solutions with linearization techniques of the fractional Harry Dym equation. *Applied Mathematics and Nonlinear Sciences*, *4*(1), 35-42.

[34] Sajid T., Sabir Z., Tanveer S., Arbi A., Altamirano GC. (2020). Upshot of radiative rotating Prandtl fluid flow over a slippery surface embedded with variable species diffusivity and multiple convective boundary conditions. *Heat Transfer*. 2020;1–21. https://doi.org/10.1002/htj.22010

[35] Dewasurendra, M., & Vajravelu, K. (2018). On the method of inverse mapping for solutions of coupled systems of nonlinear differential equations arising in nanofluid flow, heat and mass transfer. *Applied Mathematics and Nonlinear Sciences*, *3*(1), 1-14.

[36] Dusunceli, F. (2019). New exact solutions for generalized (3+ 1) shallow water-like (SWL) equation. *Applied Mathematics and Nonlinear Sciences*, *4*(2), 365-370.

[37] Yokus, A., Durur, H., & Ahmad, H. (2020). Hyperbolic type solutions for the couple Boiti-Leon-Pempinelli system. *Facta Universitatis, Series: Mathematics and Informatics*, *35*(2), 523-531.

[38] Ali, M., Shahzad, M., Sultan, F., Khan, W. A., & Shah, S. Z. H. (2020). Characteristic of heat transfer in flow of Cross nanofluid during melting process. *Applied Nanoscience*, 1-10.

[39] Pérez-García, V. M., Fitzpatrick, S., Pérez-Romasanta, L. A., Pesic, M., Schucht, P., Arana, E., & Sánchez-Gómez, P. (2016). Applied mathematics and nonlinear sciences in the war on cancer. *Applied Mathematics and Nonlinear Sciences*, *1*(2), 423-436.

[40] Shah, S. Z., Wahab, H. A., Ayub, A., Sabir, Z., Haider, A., & Shah, S. L. Higher order chemical process with heat transport of magnetized cross nanofluid over wedge geometry. Heat Transfer.

Chapter 8

Numerical Investigation of Natural Convection and Entropy Generation of Water near Density Inversion in a Cavity Having Circular and Elliptical Body

Nguyen Minh Phu and Nguyen Van Hap

Abstract

In this chapter, a water-filled square cavity with left hot wall and right cold wall was numerically investigated. The hot and cold wall temperatures are 10°C and 0°C respectively to examine the density inversion of natural convection water, i.e. water at 4°C. In the middle of the square, there are circular and elliptical bodies to study fluid–structure interaction in terms of the thermohydraulic behavior and entropy generation. 2D numerical simulation was performed using finite volume method in Ansys fluent software with the assumption of laminar flow. The simulation results are compared with benchmark data to determine reliability. The results indicate that the body insertions increase the convection heat transfer coefficients at the best heat transfer positions due to impingement heat transfer. An increase in heat transfer rate of 1.06 times is observed in the case of circular body compared to none. There are three primary eddies in the cavity with bodies, whereas the cavity without body has two primary eddies. Maximum entropy generation was found in the upper right corner of cavity mainly due to high horizontal temperature gradient. Bodies of circle and vertical ellipse have almost the same thermohydraulic and entropy generation characteristics due to the same horizontal dimension which mainly effects on the downward natural convection current. The entropy generation of cavity with circular body is 1.23 times higher than that of the cavity without body. At positions $y/L = 1$ on the hot wall and $y/L = 0.74$ on the cold wall, the convection heat transfer coefficient is close to zero due to stagnant fluid.

Keywords: natural convection, density inversion, numerical simulation, entropy generation, fluid-conic structure interaction

1. Introduction

Water is a fluid with special thermophysical properties compared to many pure substances. That is, the specific heat of water is so large that water is often thought of as a thermal storage media and the density inversion around 4°C alters the natural convection heat transfer mechanism around this temperature [1]. Early, Sasaguchi et al. [2, 3] examined water cooling with density inversion taken into account. They concluded that the position in vertical direction of the cold cylinder surface had a remarkable effect on the rate of water cooling. Tong [4] reported the effect of the aspect ratio of the cavity on the natural convection heat transfer rate of water around its maximum density. The results show that the aspect ratio of 3 enhances the density inversion. Varol et al. [5] investigated a rectangular cavity with porous media. Flow pattern and isotherms have been presented and analyzed in this study at different Rayleigh numbers. Recently, Hu et al. [6] numerically investigated circular, rectangular, and triangular cavity with inner blocks of circles, squares and triangles. They concluded that increasing the aspect ratio increases the number of vortex and that the effect of cavity's shape on heat transfer is stronger than that of the inner blocks. More recently, Cho et al. [7] simulated natural convection in a cavity with circular and elliptical objects. They confirmed that the elliptical object placed at top increased the Nusselt number by 2.1% compared to the two circular objects [8].

The above works have been done regarding the thermo-hydraulic properties of natural convection in an enclosure. Entropy generation or exergy destruction is a second consideration to comprehensively evaluate an energy system to fulfill the laws of thermodynamics [9–11]. However, the entropy generation assessment due to natural convection in an enclosure with internal objects has been little interest by researchers. Some of the works that can be found in literature are as follows. Kashani et al. [12] investigated entropy generation due to natural convection in an enclosure with vertical wavy walls. They reported that both thermal hydraulics phenomena and entropy generation are strongly influenced by density inversion. Tayebi and Chamkha [13] investigated the entropy generation of a nanofluid in a square cavity with a conducting empty cylinder. Results showed that inserting the cylinder significantly changed the heat transfer mechanism and the irreversibility of the enclosure. Li et al. [14] examines a cylinder inside an inclined enclosure. The effects of radiation and electromagnetic fields on natural convection and entropy production were considered in this study. They reported that the largest Bejan number was reached at an inclination angle of 60°.

From the literature review above, it can be seen clearly that the addition of objects into the cavity alters the thermal properties, fluid flow and entropy generation in natural convection. However, a study of natural convection of water around its maximum density with circular and elliptical objects within cavity has not been found. In this chapter, both the thermohydraulic and entropy generation mechanisms of the above problem were investigated to characterize the energy and exergy aspects of the density inversion associated with the inserts.

2. Model description and validation

Figure 1 shows the 38 mm square cavity examined in this study. Inside the cavity there are circular and elliptical bodies with basic dimensions of 18 mm and 9 mm. The cavity without body is considered as base case (Case 1). The five different conic sections of the body include circle (Case 2), vertical ellipse (Case 3), horizontal ellipse (Case 4), left inclined ellipse (Case 5), and right inclined ellipse (Case 6). Accelerate gravity vector is vertical and in the opposite direction to the y-axis to investigate natural convection in the cavity. The boundary conditions of the computational domain include the temperature T_H = 10°C in the left wall, T_C = 0°C in the right wall. The remaining walls are considered adiabatic. Natural convection currents in cavity are assumed laminar, two-dimensional and incompressible fluid [2, 3, 6, 15]. The following equations are the governing equations under current consideration:

Continuity equation

$$\frac{\partial u}{\partial x}+\frac{\partial v}{\partial y}=0 \tag{1}$$

x-momentum

$$u\frac{\partial u}{\partial x}+v\frac{\partial u}{\partial y}=-\frac{1}{\rho_0}\frac{\partial p}{\partial x}+\frac{\mu}{\rho_0}\left(\frac{\partial^2 u}{\partial x^2}+\frac{\partial^2 u}{\partial y^2}\right) \tag{2}$$

y-momentum

$$u\frac{\partial v}{\partial x}+v\frac{\partial v}{\partial y}=-\frac{1}{\rho_0}\frac{\partial p}{\partial y}+\frac{\mu}{\rho_0}\left(\frac{\partial^2 v}{\partial x^2}+\frac{\partial^2 v}{\partial y^2}\right)-g(\rho-\rho_0) \tag{3}$$

Energy equation

$$u\frac{\partial T}{\partial x}+v\frac{\partial T}{\partial y}=\frac{k}{c_p\rho_0}\left(\frac{\partial^2 T}{\partial x^2}+\frac{\partial^2 T}{\partial y^2}\right) \tag{4}$$

where ρ_0 is the reference density of water, its value of 999.8 kg/m^3 in the present study.

The density of water in the temperature range from 0 to 10°C was fitted from the EES software (F-chart software) as follows:

$$\rho=-7218.26287+106.273748T-0.517050664T^2+0.0011246302T^3 -9.25648144\times10^{-7}T^4 \tag{5}$$

where ρ is water density (kg/m^3) and T is water temperature (K).

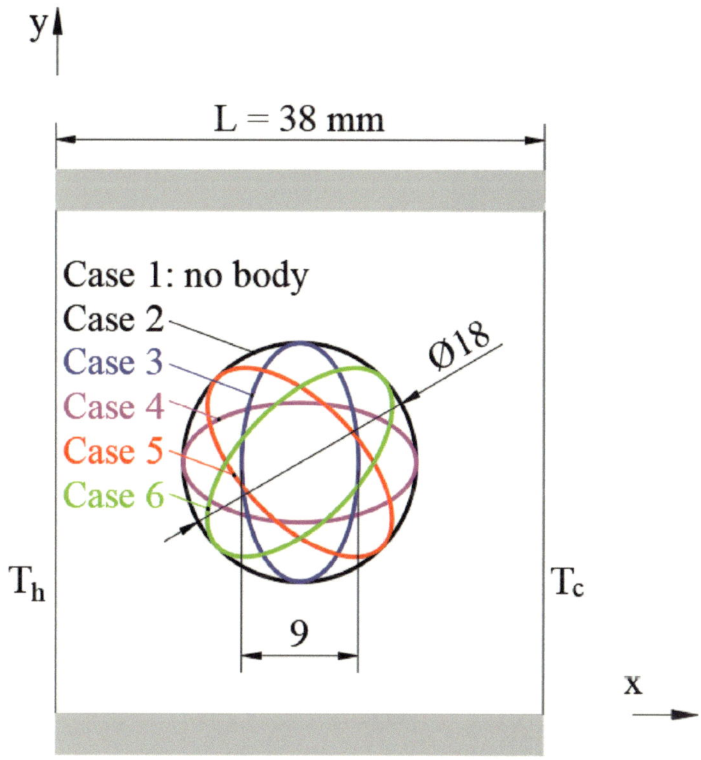

Figure 1.
Computational domain and investigated cases.

It can be observed the difficulty when dealing with the Navier–Stokes governing equation. This is due to the fact that left hand side of the momentum equations is non-linear term. Secondly, a pressure variable does not present in the continuity equation. To overcome drawbacks, a semi-implicit method was adopted in the present numerical study. The selected method is SIMPLE (Semi-Implicit Method for Pressure Linked Equations) algorithm which linearizes convection term in momentum equations and is used to couple velocity and pressure variables in Eqs. (1) and (2). Application of the SIMPLE algorithm to curved surfaces of circle and ellipse can be resolved by an interpolation between grid vertexes which are not coincident with the boundaries leading to low accuracy of the solution. Therefore, the geometric imperfection is eliminated by using a curvilinear coordinate system [16, 17].

Figure 2 shows the meshing in a typical computational domain. Refinements are enhanced for the surfaces to increase accuracy in predicting phenomena in conjunction with the boundary layer. For hybrid grids as displayed in **Figure 2** and enlarged in **Figure 3a**, the velocity and pressure variables are stored in the center of a control volume as illustrated in **Figure 3b**. Thus, these qualities at a face could be interpolated by center-stored pressure and velocity of two adjacent control volumes. The face interpolation can be treated by using the curvilinear coordinate system ξ-η as seen in **Figure 3c** where $\mathbf{n}$ is the vector normal to the face, $\mathbf{e}_\xi$ and $\mathbf{e}_\eta$ the unit vectors along ξ-η axes [18].

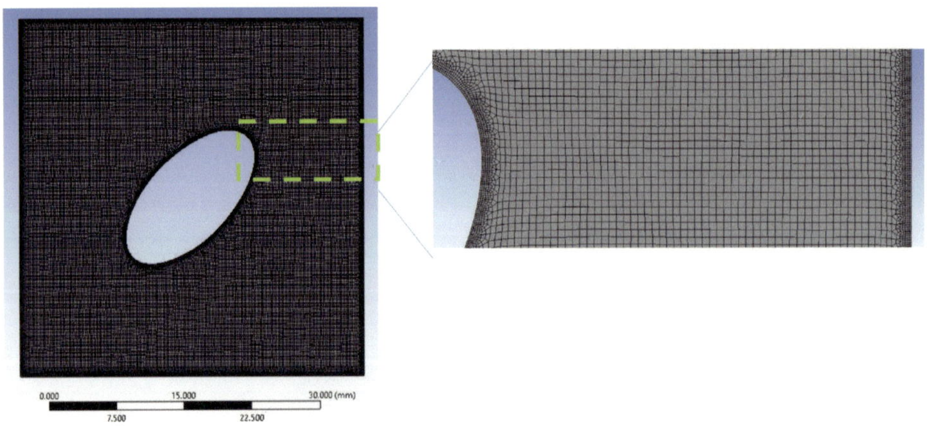

Figure 2.
Mesh generation with refinement.

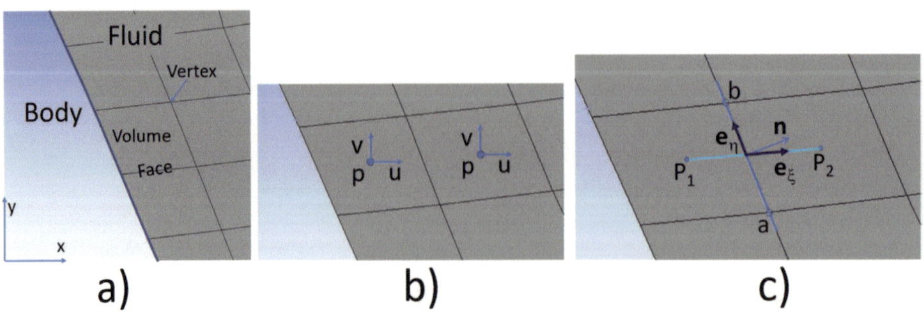

Figure 3.
Face interpolation in finite volume method. (a) Vertex, face, and volume definition. (b) Pressure and velocity components stored in center of a volume. (c) Face interpolation by means of curvilinear coordinate system ξ-η.

Grid independence check for case 1 was conducted with the number of nodes of 18,214, 28,454, 54,609 and 181,524 showing that the number of nodes 54,609 has small error compared with data in the literature and computation cost is moderate. The residuals of 1e-4 and 1e-6 are set for the momentum and energy equations, respectively. **Figure 4** displayed a comparison of velocity components in case 1 along the x-axis at the horizontal center line y = 19 mm with the number of nodes of 54,609, where X = x/L, X-velocity = uL/α, Y-velocity = vL/α, α is the thermal diffusivity. The comparison result with published data [19] showed a fairly good agreement. From these settings and confirmations, grids of about 56,000 nodes are obtained for 5 cases having a body in the middle of cavity.

The thermal and entropy generation parameters were deduced from the simulated data. The local heat transfer coefficient (HTC) at hot and cold walls is computed as:

$$h = \frac{-k\dfrac{\partial T}{\partial x}}{T_H - T_C} \tag{6}$$

where k is the conductivity of water.

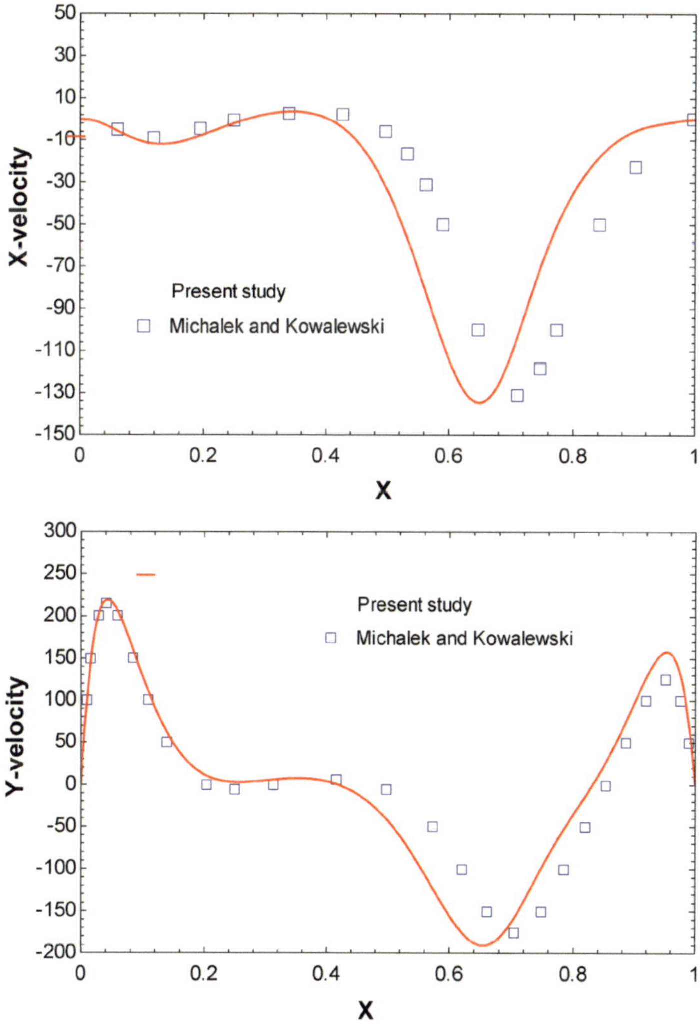

Figure 4.
Validation with published data [19].

The local entropy generation inside the domain can be estimated by [20, 21]:
The local entropy generation due to heat transfer:

$$\dot{S}_t''' = \frac{k}{T^2}\left[\left(\frac{\partial T}{\partial x}\right)^2 + \left(\frac{\partial T}{\partial y}\right)^2\right] \tag{7}$$

The local entropy generation due to fluid friction:

$$\dot{S}_v''' = \frac{\mu}{T}\left\{2\left[\left(\frac{\partial u}{\partial x}\right)^2 + \left(\frac{\partial v}{\partial y}\right)^2\right] + \left(\frac{\partial u}{\partial y} + \frac{\partial v}{\partial x}\right)^2\right\} \tag{8}$$

where μ is the water dynamic viscosity.
Average total entropy generation can be found by the volumetric integral as:

$$\overline{\dot{S}}''' = \frac{1}{V}\int_V \left(\dot{S}_t''' + \dot{S}_v'''\right) dV \tag{9}$$

where V is the volume of computational domain.

3. Results and discussion

Figure 5 shows the velocity distribution in square cavity. Fluid–structure interactions can be clearly observed for six cases. The contour in the **Figure 5** exhibited high-speed upward flow near the heat exchange surfaces and downflow in the middle due to the density difference. The maximum natural convection velocity is about 0.9 mm/s. Case 1 (without body) showed a lightly higher velocity than others due to motion obstruction of the body. It can be seen that the flow pattern of case 2 (circle) and case 4 (horizontal ellipse) is nearly the same. This is because the horizontal size of the two bodies is the same (18 mm). This is the main length that affects downward stream.

The temperature field inside the cavity can be seen in **Figure 6**. The temperature stratifications are clearly visible in the upper half cavity. The upper left corner has

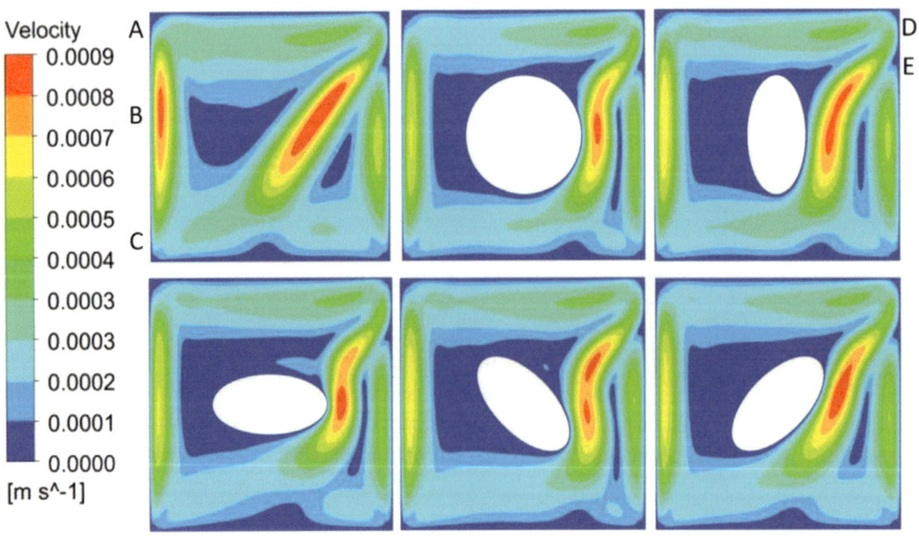

Figure 5.
Velocity magnitude.

Figure 6.
Isotherms.

high temperature due to the fluid that receives heat from the hot surface and rises. There is a region with relatively low temperature near the cold surface due to the interference between the downstream and the upstream causing stagnant fluid (point E on **Figure 5**).

The flow pattern in the cavity can be observed through the streamlines as shown in **Figure 7**. It is clear that the number of vortices increases from 2 to 3 for the absence of body to its existence. The third swirl located on the upper right side of the body. The one more vortex is formed by the separation of the downstream flow by the bodies. The eddies in the body-inserted cases are smaller than those of the base case due to the occupation of the bodies.

Figure 8 shows the total entropy generation distribution. It can be seen that the largest entropy generation is the upper right corner, the second largest is the lower left corner. Case 2 had the largest entropy generation and case 1 indicated the smallest

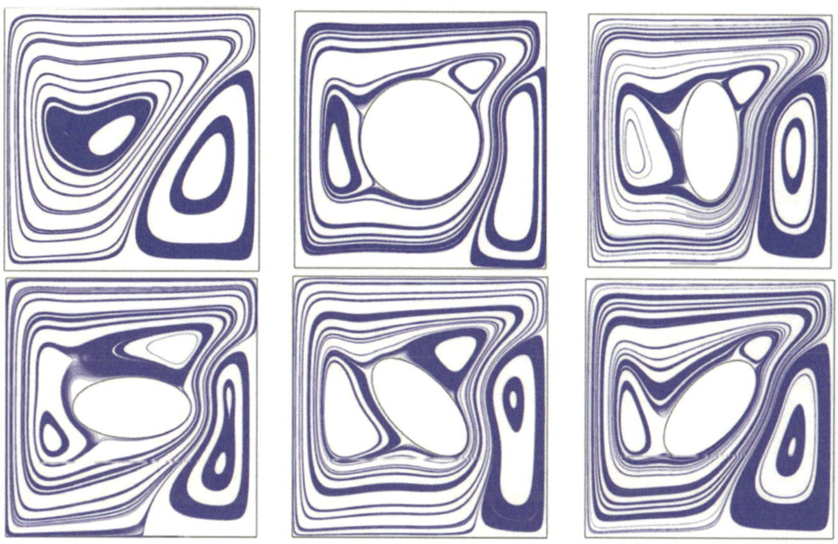

Figure 7.
Streamlines.

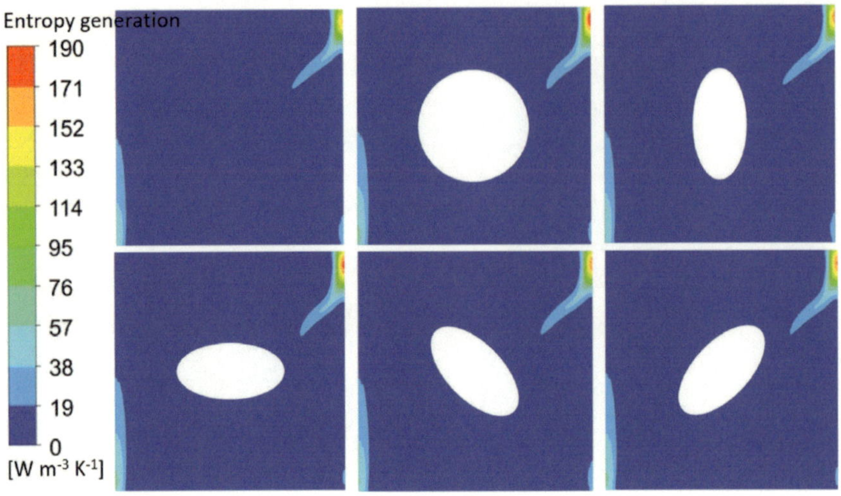

Figure 8.
Total entropy generation.

one. The maximum entropy generation is about 190 W/m^3-K. The entropy generation distribution is quite similar to that of Kashani et al. [12]. The laminar flow and the low velocity lead to small entropy generation due to friction. Entropy generation in the cavity is prevailed by heat transfer. **Figure 9** shows the temperature gradients in case 2 to clarify the effect of the components. Combining **Figures 6** and **9** shows that the horizontal temperature variation is large in the upper right corner. Since downward flow is accompanied by strong heat exchange, the temperature gradient in x direction has considerably high magnitude. The second largest entropy generation is close to the stagnant point on cold wall caused by a large temperature gradient in y-direction. **Figure 10** compares the mean entropy generation for the cases under consideration. The largest and smallest average entropy generations are 3.7 W/ m^3-K and 3 W/m^3-K respectively, corresponding to the entropy generation in case 2 of 1.23 times higher than the base case. The other cases have the average entropy generation of 3.35 W/m^3-K.

Figure 11 reported the local heat transfer coefficient (HTC) on hot and cold surfaces. The largest HTC for the hot surface finds near the bottom wall due to reverse flow causing impingement heat transfer. The local heat transfer coefficient decreased

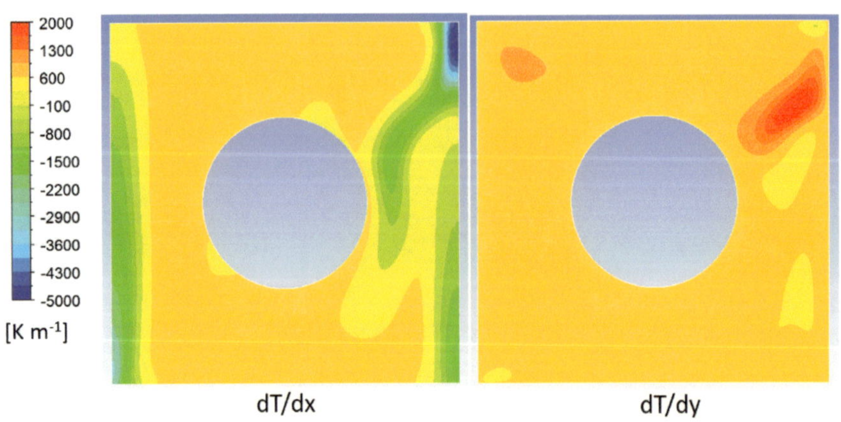

Figure 9.
Temperature gradients in case 2.

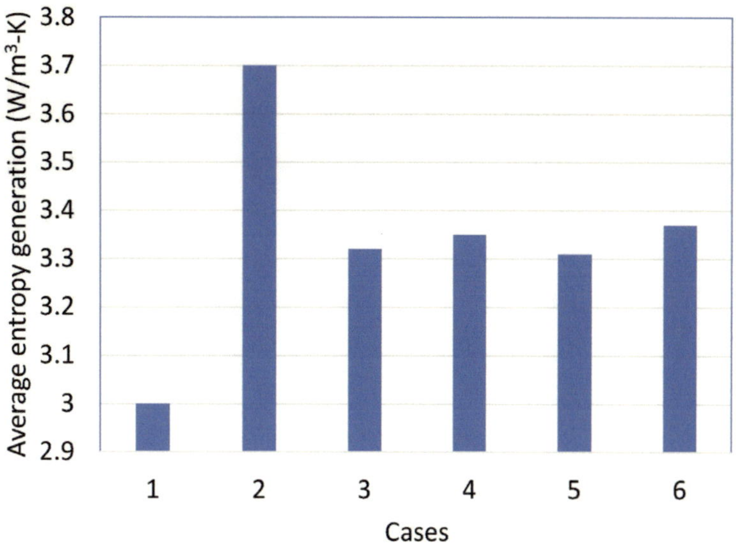

Figure 10.
Average entropy generation.

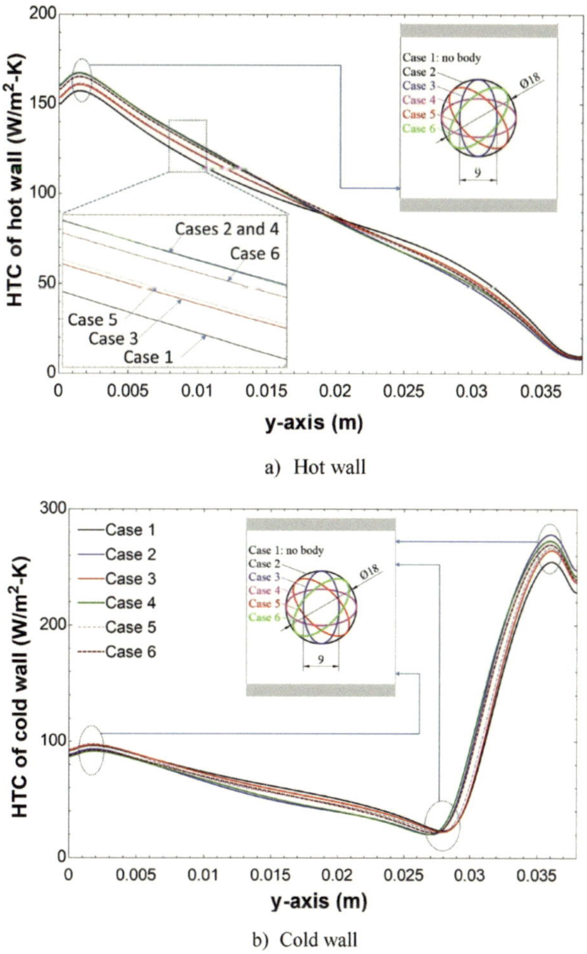

Figure 11.
Local heat transfer coefficient. (a) Hot wall; (b) cold wall.

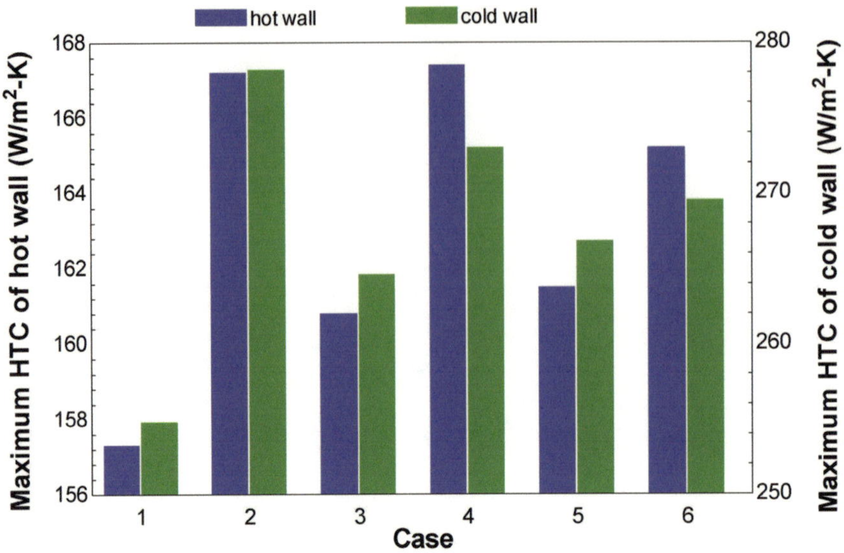

Figure 12.
Comparison of maximum heat transfer coefficient.

with increase in height. This is due to the fact that the fluid increases the temperature leading to reduce the temperature difference between the hot wall and the adjacent fluid. At the top of the hot surface (y = L) the fluid was not moving. Therefore, the convection heat transfer coefficient approaches zero. Body insertion increased the HTC at the best heat transfer position ($y \approx 0$). This is because the flow is guided by the body and acts perpendicular to the hot surface as seen in **Figure 5** (point C). In this region, circular body gives the highest heat transfer coefficient because its shape plays a role as a guide vane. Similar phenomena can be observed for HTC of the cold side as shown in **Figure 11b**. At position y = 35 mm HTC is the largest due to impingement heat transfer (see more point D on **Figure 5**). At y = 28 mm (y/L = 0.74), the HTC is approximately zero due to the stagnant fluid as explained above. When y increases from 0 to 28 mm, the HTC decreases due to the decrease in temperature gradient in x-direction. At the best heat transfer position of the cold surface, i.e. y = 35 mm, we can see that the circular body gives the highest HTC and the base case results in the smallest HTC.

Figure 12 compares the maximum heat transfer coefficient on hot ($y \approx 1.5$ mm) and cold ($y \approx 35$ mm) walls. It can be seen that the HTC of a circle-inserted cavity is 1.06 times ($\approx 167/157$) higher than that of the base case (Case 1). Once again, we can see that the HTC of cases 2 and 4 is quite identical. Because circular body and elliptical body have the same horizontal dimension (18 mm). The natural convection flow in an enclosure is prevailed by width of a body due to upwelling and downwelling plumes. Among the inserted cavity, case 3 (vertical ellipse) has the smallest HTC due to its smallest horizontal length.

4. Conclusions

The numerical study of the natural convection of water around its maximum density was carried out in this chapter. Fluid-conic structure interaction and natural convection heat transfer were presented and analyzed. The characteristics of temperature distribution, water fluid flow, local entropy generation and local convection heat transfer coefficient were investigated. The main findings from the chapter are as follows:

1. The body insertions prevent the middle downstream causing a slight decrease in fluid velocity.

2. The flow pattern of the circular body and the horizontal elliptical body are nearly identical.

3. The bodies inserted into cavity increase one more primary vortex because of the downward flow separated by bodies.

4. The circular body has the largest entropy generation, the cavity without body yields the smallest entropy generation.

5. Maximum entropy generation occurs at the upper right corner of the cavity. This comes from the huge x-direction temperature variation in this region.

6. Inserts increase the greatest heat transfer coefficient on hot and cold walls. In which the biggest increase is circular body followed by the horizontal elliptical body.

Author details

Nguyen Minh Phu[1,2]* and Nguyen Van Hap[1,2]

1 Faculty of Mechanical Engineering, Ho Chi Minh City University of Technology (HCMUT), Ho Chi Minh City, Viet Nam

2 Vietnam National University, Ho Chi Minh City, Vietnam

*Address all correspondence to: nmphu@hcmut.edu.vn

References

[1] Nguyen MP, Lee G-S. Ice Formation on the Outer Surface of a Vertical Tube with Inside Refrigerant Boiling. Transactions of the Korean Society of Mechanical Engineers B. 2011;**35**(2):129-135 https://doi.org/10.3795/ksme-b.2011.35.2.129

[2] Sasaguchi, K., Kusano, K., Kitagawa, H., & Kuwabara, K. (1997). EFFECT OF DENSITY INVERSION ON COOLING OF WATER AROUND A CYLINDER IN A RECTANGULAR CAVITY. Numerical Heat Transfer, Part A: Applications, 32(2), 131-148. https://doi.org/10.1080/10407789708913884

[3] Sasaguchi, K., Kuwabara, K., Kusano, K., & Kitagawa, H. (1998). Transient cooling of water around a cylinder in a rectangular cavity—a numerical analysis of the effect of the position of the cylinder. International Journal of Heat and Mass Transfer, 41(20), 3149-3156. https://doi.org/10.1016/s0017-9310(98)00064-7

[4] Tong, W. (1999). Aspect ratio effect on natural convection in water near its density maximum temperature. International Journal of Heat and Fluid Flow, 20(6), 624-633. https://doi.org/10.1016/s0142-727x(99)00027-2

[5] Varol, Y., Oztop, H. F., Mobedi, M., & Pop, I. (2010). Visualization of heat flow using Bejan's heatline due to natural convection of water near 4°C in thick walled porous cavity. International Journal of Heat and Mass Transfer, 53(9-10), 1691-1698. https://doi.org/10.1016/j.ijheatmasstransfer.2010.01.020

[6] Hu, Y.-P., Li, Y.-R., & Wu, C.-M. (2014). Comparison investigation on natural convection of cold water near its density maximum in annular enclosures with complex configurations. International Journal of Heat and Mass Transfer, 72, 572-584. https://doi.org/10.1016/j.ijheatmasstransfer.2014.01.047

[7] Cho, H. W., Park, Y. G., & Ha, M. Y. (2018). The natural convection in a square enclosure with two hot inner cylinders, Part I: The effect of one elliptical cylinder with various aspect ratios in a vertical array. International Journal of Heat and Mass Transfer, 125, 815-827. https://doi.org/10.1016/j.ijheatmasstransfer.2018.04.141

[8] Phu, N. M., Bao, T. T., Hung, H. N., Tu, N. T., & Van Hap, N. (2020). Analytical predictions of exergoeconomic performance of a solar air heater with surface roughness of metal waste. Journal of Thermal Analysis and Calorimetry. https://doi.org/10.1007/s10973-020-09787-5

[9] Phu, N. M. (2019). Overall Optimization and Exergy Analysis of an Air Conditioning System Using a Series-Series Counterflow Arrangement of Water Chillers. International Journal of Air-Conditioning and Refrigeration, 27(04), 1950034. https://doi.org/10.1142/s2010132519500342

[10] Nguyen, P. M. (2016). Energy and exergy estimation for a combined cycle of solid CO2 production and NH3-H2O single effect absorption chiller. Science and Technology Development Journal, 19(1), 61-69. https://doi.org/10.32508/stdj.v19i1.611

[11] Luan, N. T., & Phu, N. M. (2020). Thermohydraulic correlations and exergy analysis of a solar air heater duct with inclined baffles. Case Studies in Thermal Engineering, 21, 100672. https://doi.org/10.1016/j.csite.2020.100672

[12] Kashani, S., Ranjbar, A. A., Mastiani, M., & Mirzaei, H. (2014). Entropy generation and natural convection of nanoparticle-water mixture (nanofluid)

near water density inversion in an enclosure with various patterns of vertical wavy walls. Applied Mathematics and Computation, 226, 180-193. https://doi.org/10.1016/j.amc.2013.10.054

[13] Tayebi, T., & Chamkha, A. J. (2019). Entropy generation analysis due to MHD natural convection flow in a cavity occupied with hybrid nanofluid and equipped with a conducting hollow cylinder. Journal of Thermal Analysis and Calorimetry, 139(3), 2165-2179. https://doi.org/10.1007/s10973-019-08651-5

[14] Li, Z., Hussein, A. K., Younis, O., Afrand, M., & Feng, S. (2020). Natural convection and entropy generation of a nanofluid around a circular baffle inside an inclined square cavity under thermal radiation and magnetic field effects. International Communications in Heat and Mass Transfer, 116, 104650. https://doi.org/10.1016/j.icheatmasstransfer.2020.104650

[15] Dalal, A., & Das, M. K. (2005). Laminar natural convection in an inclined complicated cavity with spatially variable wall temperature. International Journal of Heat and Mass Transfer, 48(14), 2986-3007. https://doi.org/10.1016/j.ijheatmasstransfer.2004.07.050

[16] Rhie, C. M., & Chow, W. L. (1983). Numerical study of the turbulent flow past an airfoil with trailing edge separation. AIAA Journal, 21(11), 1525-1532. https://doi.org/10.2514/3.8284

[17] Versteeg, H. K., & Malalasekera, W. (2007). An Introduction to Computational Fluid Dynamics - The finite volume method, Second Edition, Pearson.

[18] Jiannan Tan (2010). A study of solving Navier-Stokes equations with a finite volume method based on polygonal unstructured grids and the computational analysis of ground vehicle aerodynamics, PhD. Dissertation, Texas Tech University.

[19] Michalek, T., Kowalewski, T. A., & Sarler, B. (2005). Natural convection for anomalous density variation of water: numerical benchmark. Progress in Computational Fluid Dynamics, An International Journal, 5(3/4/5), 158. https://doi.org/10.1504/pcfd.2005.006751

[20] Phu, N. M., & Van Hap, N. (2020). Performance Evaluation of a Solar Air Heater Roughened with Conic-Curve Profile Ribs Based on Efficiencies and Entropy Generation. Arabian Journal for Science and Engineering. https://doi.org/10.1007/s13369-020-04676-3

[21] Phu, N. M., & Luan, N. T. (2020). A Review of Energy and Exergy Analyses of a Roughened Solar Air Heater. Journal of Advanced Research in Fluid Mechanics and Thermal Sciences, 77, 160-175.https://doi.org/10.37934/arfmts.77.2.160175

Chapter 9

Seismic Response Characteristics of RCC Dams Considering Fluid-Structure Interaction of Dam-Reservoir System

Khaled Ghaedi, Farzad Hejazi, Meisam Gordan, Ahad Javanmardi, Hamed Khatibi and Ali Joharchi

Abstract

In analysis of different types of dams, i.e. arch, gravity, rockfill and Roller Compacted Concrete (RCC) dams, the effect of hydrodynamic water pressure as an effective factor must seriously be taken into consideration. In present study, the hydrodynamic effect is precisely deliberated in RCC dams and compared to hydrostatic pressure effect. For this purpose, Kinta RCC dam in Malaysia is selected and 2D finite element (FE) model of the dam is performed. The Lagrangian approach is used to solve the dam-reservoir interaction, fluid–structure interaction (FSI), and in order to evaluate the crack pattern, Concrete Damaged Plasticity (CDP) model is implemented. Comparisons show that hydrodynamic pressure significantly changes the dam behaviour under seismic excitations. Moreover, the hydrodynamic effect modifies the deformation shape of the dam during the ground motions, however, it increases the magnitudes of the developed stresses causing more extensive tension crack damages mostly in the heel and upstream zones of the dam.

Keywords: fluid–structure interaction, hydrodynamic pressure, earthquake, concrete damaged plasticity, Kinta dam

1. Introduction

Earthquake as an unpredictable event [1] is one of the main concerns of structural engineers. However, to protect civil structures such as buildings and bridges against ground motions several approaches have so far been used [2–9], but more attention must be payed to construct water control structures such as dams, where the water weight effect combined with earthquake force increases the danger of structural destruction. Recently, application of RCC technology in dam construction was launched, early 2002. This technology provides some advantages for dam engineers in terms of equipment, manpower, construction speed and cost. Analysis of gravity dams subjected to earthquake excitations considering different aspects of analysis including interactions, boundary conditions, reservoir length and height have been addressed by many researchers [10–15]. The difference between constitutive relationship of RCC and conventional concrete invites the investigators to study different aspects of analysis of the RCC dams such as dam-foundation,

 IntechOpen

dam-reservoir and dam-reservoir-foundation interaction. For instance, Fenves and Chopra [16] presented a simplified method to evaluate the response of concrete gravity dams utilizing fundamental vibration mode considering dam-reservoir with impounded water interaction and dam-foundation interaction. Malkar [17] investigated the seismic responses of several gravity dams considering different heights by means of Automatic Dynamic Incremental Nonlinear Analysis (ADINA) code. The crack propagation model was also investigated using fracture criterion of tensile stresses. Ayari [18] used the effective simulation of fracture mechanics and discrete crack (DC) closure under transient dynamic circumstances to develop new models in order to investigate the crack propagation of concrete gravity dams. To this end, the Koyna gravity dam was selected to be analysed. Espandar and Lotfi [19] applied bidirectional accelerations to the Shahid Rajaee arch dam to investigate the nonlinear seismic response of the dam using the nonlinear techniques of continuum mechanics i.e. elasto-plastic and non-orthogonal smeared crack (NOSC) models. Later on, the accuracy of the models was compared to experimental results. It was concluded that, the NOSC model was much better than the linear elastic analysis to evaluate the dam stresses. Lotfi and Espandar [20] developed a special FE program using combination of DC and NOSC methods, known as DC-NOSC technique, which was employed to examine the nonlinear behaviour of an arc dam. It was found that the results of the DC-NOSC model were more reliable compared to DC and NOSC method used alone. Akkose et al. [21] examined the response of arc dams considering the effect of different water levels. The yield criterion of Drucker–Prager was implemented to idealize the concrete dam in nonlinear analysis. The reservoir water was modelled according to the Lagrangian method. Akkose and Simsek [22] investigated the influence of both near- and far-fault excitations on the nonlinear response of a gravity dam considering dam-reservoir-foundation-sediment interaction. Kartal [23] imposed a three-directional earthquake to a RCC dam in order to study the response of the dam considering material and geometry. The effect of reservoir water was inspected using the Lagrangian method through fluid FEM. Zhang et al. [24] investigated the effect of strong after-shocks and their potential to damage the concrete gravity dams. Hence, the hardening behaviour was taken into account for material properties of the dams to deliberate the crack propagation. Ghaedi et al. [25] studied the influence of flexible foundations on seismic response of RCC dams using FEM considering hydrodynamic pressure of the reservoir water. Wang et al. [26] studied the correlation between period of ground excitations and damage severity in gravity dams using twenty ground motion records with different time frames. They showed that, durations of the ground motions had no remarkable effect on damage levels. Ghaedi et al. [27] investigated the effect of openings (galleries) considering different sizes and shapes on dynamic response of RCC dams. The results showed that, RCC dams with circular openings behave better compared to other geometric shapes such assquare and octagon. Wang and Jia [28] made an attempt to propose a new approach to test and design hydraulic fracturing of high concrete gravity dams with more than 200 m height. For that purpose, a cylindrical sample with embedded crack was prepared to model the hydraulic fracture of concrete gravity dam heel, while the sample was subjected to a uniaxial load at both ends of the sample. Wang et al. [29] constructed a large-scale concrete-rockfill combination dam (CRCD) and crucial factors such as deformation, acceleration time history, slope failure and dynamic earth pressure were examined to investigate the dynamic performance of the CRCD.

Based on the above literature, it is a crucial issue to take hydrodynamic reservoir pressure effect into account in order to investigate the behaviour of RCC dams under seismic motions. To aid the aim, an attempt is made to investigate the nonlinear behaviour of a RCC dam subjected to earthquake excitations considering the hydrodynamic water pressure effect through the Lagrangian approach. For this purpose, Kinta RCC dam located in Malaysia is chosen as a case study and a two-dimensional Finite Element Model (FEM) is implemented via ABAQUS software. In order to predict the crack propagation, the Concrete Damaged Plasticity (CDP) model is used. In addition, the model change technique is used to model the RCC layers of the dam.

2. Kinta RCC dam (case study)

The Kinta RCC dam location in the city of Ipoh, 205 km north from Kuala Lumpur. The Kinta dam is the first constructed RCC dam in Malaysia, as shown in **Figure 1**, and for the first time RCC technology had been utilized in the world to construct the steps of spillway. To build the dam, RCC with zero-slump was used while several slim horizontal layers were compacted using vibratory rollers. In the present study, the Kinta RCC dam is chosen as a case study to evaluate the seismic behaviour of the dam when the effect of reservoir water pressure is taken into consideration. The structural geometry of the deepest section of the Kinta RCC dam used in the study is depicted in **Figure 2** [31] (see also [27]).

As shown in **Figure 2**, the dam comprises three sections including the dam body which is surrounded by Conventional Vibrated Concrete (CVC) upstream and downstream face as well as CVC foundation.

3. Finite element model (FEM)

For eismic analysis of the Kinta RCC dam, the dam is accurately modelled using FE software, ABAQUS. To discretize the dam body as well as CVCs, a 2D isoparametric elements with four nodes bilinear plane stress quadrilateral, reduced integration and hourglass control is implemented. Besides, to discrete the reservoir

Figure 1.
Location of Kinta RCC dam.

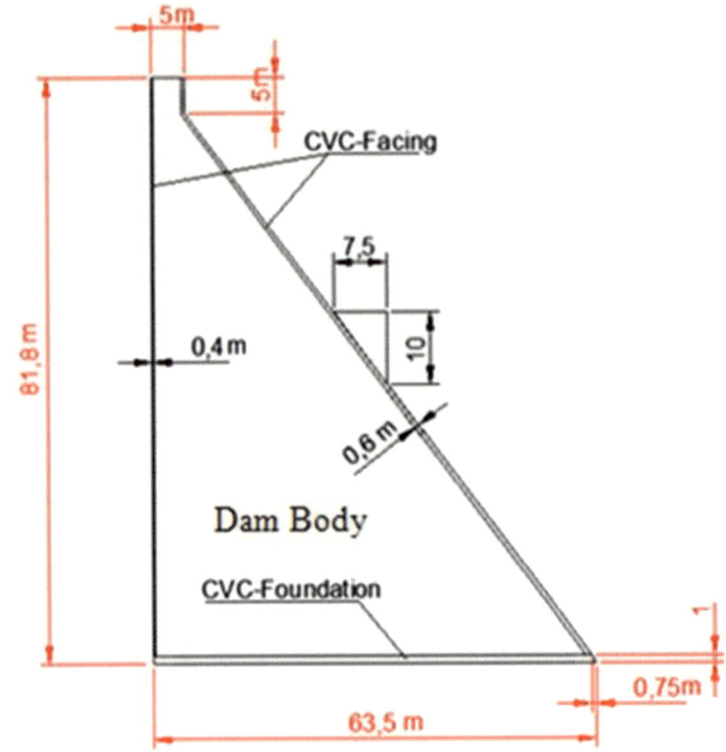

Figure 2.
Geometry of the Kinta RCC dam [30].

Block	No. of nodes	No. of elements
Dam Body	609	560
CVC Upstream Facing	56	27
CVC Downstream Facing	58	26
CVC Foundation	42	19
Reservoir Water	551	504

Table 1.
Finite element discretization of the Kinta RCC dam.

water, a 2D acoustic quadrilateral FE with four nodes linear is conducted. The details of the dam and CVCs discretization are indicated in **Table 1**. **Figure 3** shows the FEM of the Kinta dam. Furthermore, the material properties are used for modelling of the dam as demonstrated in **Table 2** [32]. The tensile strength is characterized as 10% of the compressive strength [33]. The density, ρ, and the bulk modulus, K_w, for the reservoir water are taken as 1000 Kg/m^3 and 2107 MPa, respectively.

3.1 Fluid–structure interaction (FSI)

Three basic methods are frequently utilized to solve the fluid–structure interactions using the FEM, i.e. Eulerian method, Westergaard method, and Lagrangian method. In Eulerian method, translations (displacements) are considered as variables in structure, whereas, variables of fluid are pressures. Because the structure and

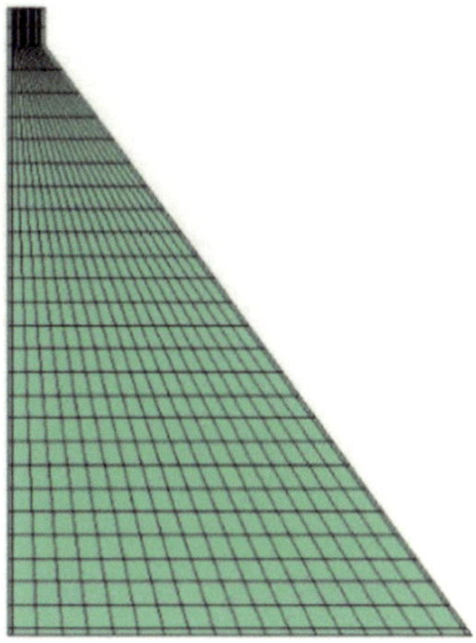

Figure 3.
FEM of the Kinta RCC dam.

Material property	Young modulus (MPa)	Poisson ratio	Density (Kg/m^3)	$\sigma_{cu}\sigma_{cu}$ (MPa) Ultimate compressive stress	σ_{tu} (MPa) Ultimate tensile stress
RCC DAM BODY	23000	0.2	2386	20	2.5
CVC-FACING	32000	0.2	2352	40	5
CVC-FOUNDATION	23000	0.2	2325	20	2.5

Table 2.
Material properties used in the present study.

fluid variables are not same in this method, a special purpose computer is required to solve the coupled systems. In Westergaard method, added mass, connected to the structure, is used to represent fluid–structure system. In Lagrangian method, variables in both fluid and structure are similar. Therefore, in this method the equilibrium as well as the compatibilities are repeatedly converged along the interface nodes. This is displacement-based FE for fluid elements that is quite appropriate because it is not required unusual interface equations and it can be performed via general-purpose computers [27]. As a result, in the present study in order to investigate the effect of water pressure, the Lagrangian method was used.

3.2 Finite element equation

The FSI has to be considered for purpose of the nonlinear analysis of the dam during earthquake excitation. Therefore, the finite element discretization of the differential equation defines the displacement of the dam structure as below:

$$M_s\ddot{u} + C_s\dot{u} + K_s = F_g + F_P \tag{1}$$

where the M_s is mass and C_s and K_s are damping and stiffness, respectively. $\ddot{u}, \dot{u}$ and $u\acute{u}, \acute{u}\wedge u$ are the relative acceleration, velocity and displacement of the dam with respect to the time, t. In addition, F_g and F_P are the force and extra force vectors described as:

$$F_g = -M_s I \ddot{u}_g(t) \tag{2}$$

and

$$F_P = \text{QP} \tag{3}$$

In which, I is the influence vector and F_P is the hydrodynamic force acting towards the dam at the upstream face. This force is a function of unidentified parameters of the nodal pressure vector of the water, P, through the transformation matrix, Q, that is determined as:

$$Q = \int N_U^T n N_p \text{dr} \tag{4}$$

where N_p is the shape function of the pressure fields and N_U is the nodal displacement of the dam. r_1 is the dam-reservoir interface and n is the unit normal vector. This explanation is owing to the discretization of the boundary conditions. The energy dissipation inside the dam is also categorized by the Rayleigh damping matrix that is written as below equation:

$$C_s = \alpha M_s + \beta K_s \tag{5}$$

in which α and β are the Rayleigh damping parameters of the first and last mode of vibrations. It is commonly agreed that, damping ratios of the dams have a range between approximately 2–5%. For all mode of vibrations, herein, the properties of material damping are adjusted to be as 5% fraction of the critical damping [27] for the first mode of the dam vibration during the dynamic analysis. In the present study, the damping ratio (5%) is tuned for the entire dam under free vibration. Accordance with this analysis, the natural frequencies of the dam are determined and ranged between ω_1= 9.571 Hz and ω_2= 51.238 Hz for the first and last mode of the dam vibrations. Consequently, by taking the natural frequencies into consideration, the values of the Rayleigh damping parameters are obtained as α = 0.806 and β = 0.00164 in the dynamic analysis of the Kinta RCC dam [34] in this study.

3.3 Coupled dam-reservoir (FSI) equations

During earthquake motions, the dam interacts with reservoir. Thus, the hydrodynamic pressure effect owing to the reservoir water and its interaction with the dam has to be taken into account. Consequently, to deliberate the reservoir hydrodynamic pressure, the force vector, F_q, due to the acceleration ($\ddot{u}$) is imposed to the upstream side of the dam. Accordingly, the force vector can be determined as:

$$F_q = -\int_{r_1} N_p^T \rho \ddot{u}.\text{ndr} \tag{6}$$

Substituting the acceleration vector into the nodal vector, that is $\ddot{u} = N_u \ddot{U}$ provides the transposed matrix of the Q^T in Eq. (4). Multiplying the transposed matrix by reservoir water density, ρ, presents the equation below:

$$K_F P = -\rho Q^T \ddot{U} \tag{7}$$

In which, $K_F = \left[G^{-1}\right]^T [A] \left[G^{-1}\right]$, A is the linear element matrix, G is the basic solution of Laplace's equation or Green function and $K_F P = F_q$. Then, the pressure vector can be concluded as:

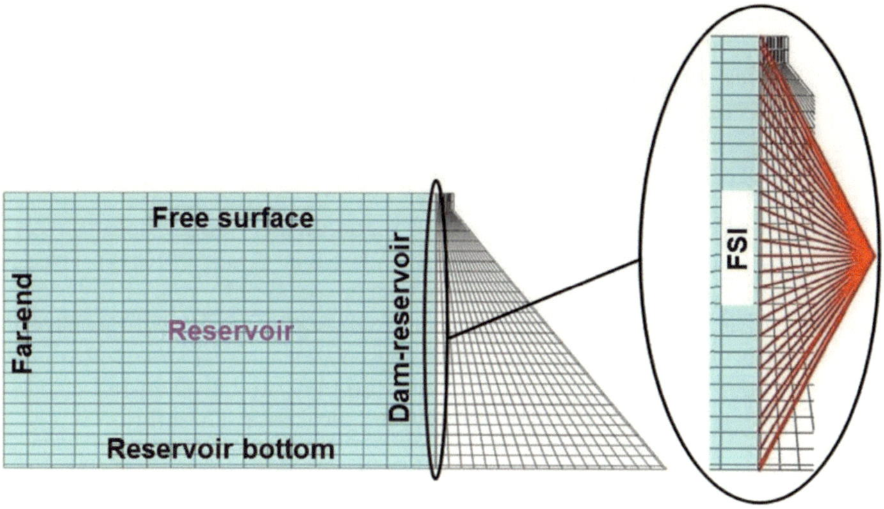

Figure 4.
Boundary conditions of the Kinta dam-reservoir system.

$$P = -\rho K_F^{-1} Q^T \ddot{U} \tag{8}$$

Substituting Eq. (8) into Eq. (3) gives the dynamic equation for displacement of the structure as below:

$$\left[M_s + \rho \mathrm{Q} \mathrm{K}_F^{-1} Q^T\right] \ddot{U} + C_s \ddot{U} + K_s U = \mathrm{F} \tag{9}$$

This is the renowned equation in FEM for solving fluid–structure systems. In this study, the reservoir water pressure is subjected to the boundary conditions at: (1) reservoir bottom (2) free surface of the reservoir (3) reservoir far-end and (4) upstream side of the dam as described by [27] and as depicted in **Figure 4**. For FSI, node to node interaction method is used in order to assess the real behaviour of the RCC dam during seismic excitation.

4. Concrete damaged plasticity (CDP)

The linear assumption may not be fit to investigate RCC dams subjected to dynamic motions [35–38]. To date, many techniques such as smeared crack model, isotropic and anisotropic damage model have proposed to investigate the constitutive model of concrete materials and their complex mechanical reaction while subjected to ground excitations. In this regard, a basic model known as plastic-damage model was proposed [39] and adapted [40]. The nonlinear performance of each synthetic material in a multiphase compounded material is commonly expressed by the CDP model and it can be used to examine the cracking. Besides, the CDP model factorizes the uniaxial compressive and tensile strength into two parts to describe the permanent degradation of stiffness and deformation. The plastic-damage model considers two major failure mechanisms for concrete materials in both compression and tension conditions, namely crushing and cracking, respectively.

The incremental theory of plasticity splits the strain tensor (ε) into two sections, i.e. the elastic strain (ε^e) and the plastic strain (ε^p) in which the equation of linear elasticity can be written as:

$$\varepsilon = \varepsilon^e + \varepsilon^p \tag{10}$$

The variables $\{\varepsilon^e, \varepsilon^p, \kappa\}$ are time-dependent. By using these parameters, the below equation can express the stress tensor:

$$\sigma = (1-d)\bar{\sigma} = (1-d)E_0(\varepsilon - \varepsilon^p) \tag{11}$$

Where $d = d(\kappa)$ is the scalar stiffness degradation which ranges from 0 (undamaged) to 1 (fully damaged); E_0 is the undamaged elastic stiffness. The material failure mechanism associates with damage, thus, reduction of the elastic stiffness is considered as a function of the internal variable (κ) i.e. compressive and tensile variables; $\kappa = (\kappa_c, \kappa_t)$. The damage functions, tension (d_t) and compression (d_c), are considered as the nonlinear functions and they can be calculated using uniaxial compressive response alongside with practical data. Therefore, the effective stress can be defined as:

$$\bar{\sigma} = (\sigma/1 - d) = E_0(\varepsilon - \varepsilon^p) \tag{12}$$

5. Loading on the dam

5.1 Hydrostatic load

Dams are usually constructed for the purpose of raising the water level of waterways on the upstream face. The rising water results hydrostatic pressure and leads the structure to slip horizontally and overturn about the toe or bottom edge of the downstream side. This pressure acts as a linear force along the dam height. In this study, the hydrostatic pressure is considered as a perpendicular force to the upstream surface so that the hydrostatic pressure is increased linearly along the dam height from zero at the free surface to 802458 Pa at the base level of the Kinta dam in the upstream side.

5.2 Hydrodynamic load

Vertical acceleration of an earthquake decreases the unit weight of material of concrete dams and the horizontal component acting on the stored water results an immediate pressure increase in the reservoir water. As a result, dam accelerates in direction of the reservoir when the water attempts to prevent the motion because of its inertia. The further pressure applied with this trend is called hydrodynamic pressure. In the present study, a reservoir with a height of 81.8 m, as displayed in **Figure 5**, is modelled to validate the hydrodynamic force at the interface of the dam-reservoir. An attempt is made to consider the hydrodynamic pressure by modeling of the accumulated water at the upstream face of the dam body. The foundation is assumed to be rigid and the impounded water in the reservoir is considered as the compressible fluid during the analysis, while the sediment absorption effect of the reservoir bottom is not deliberated.

5.3 Seismic Loading

To investigate the effect of earthquake motions on the Kinta RCC dam, Koyna earthquake records (India, 1967) with peak ground acceleration (PGA) of 0.47 g in horizontal (longitudinal) direction and 0.31 g in vertical (transverse) direction were

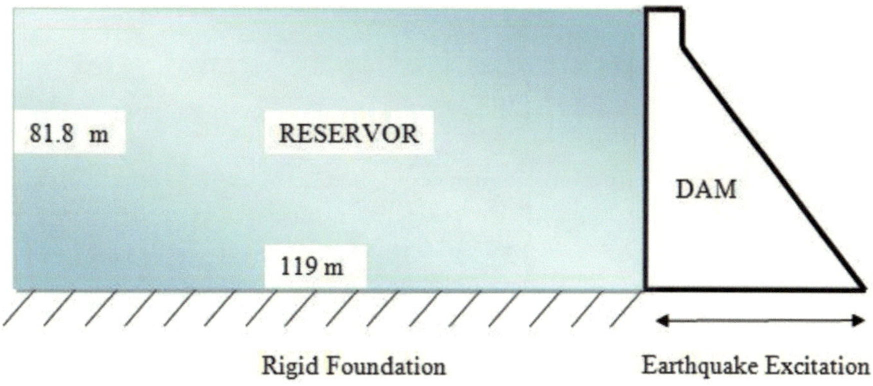

Figure 5.
Dam-reservoir model used to study hydrodynamic pressure on the Kinta RCC dam.

applied to the Kinta dam, as shown in **Figure 6**. The period of the applied earthquake is selected for 10 seconds with defining step time of 0.01 second for each interval during excitations.

6. Results and discussions

Prior to the interpretation of results and discussions, it is necessary to define the topmost node and the lowest node of the modelled dam at the upstream face in order to investigate the maximum relative acceleration and displacement response of the dam. In other words, to obtain the maximum relative acceleration and displacement response, the difference of the maximum acceleration and displacement responses of the topmost node and the lowest node (nodal acceleration and nodal displacement) has to be taken into account. For this purpose, the nodes located at the dam crest and heel zone is selected as shown in **Figure 7**.

6.1 Displacement response

The relative horizontal and vertical displacements of different dams have been reported by many researchers. Herein, the maximum relative horizontal displacement of the dam crest with hydrostatic and hydrodynamic pressure effect is investigated and indicated in **Figure 8**. As it is obvious in the figure, the seismic response of the relative horizontal displacement of the dam crest is 2.32 cm at 4.02 second due to the hydrostatic pressure and 2.8 cm at 3.689 second due to the hydrodynamic water pressure. The occurred horizontal displacements for both cases at the mentioned seconds are also demonstrated in **Figures 10(a)** and **11(a)** as the displacement contour. Therefore by subtracting the values in those figures, the relative

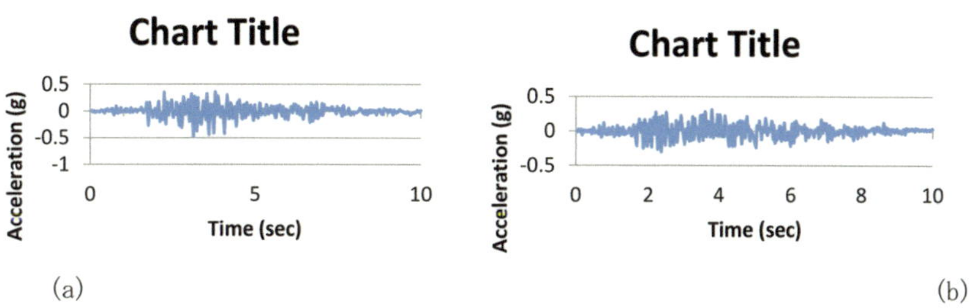

Figure 6.
(a) Horizontal and (b) Vertical acceleration of the Koyna excitations (India, 1967).

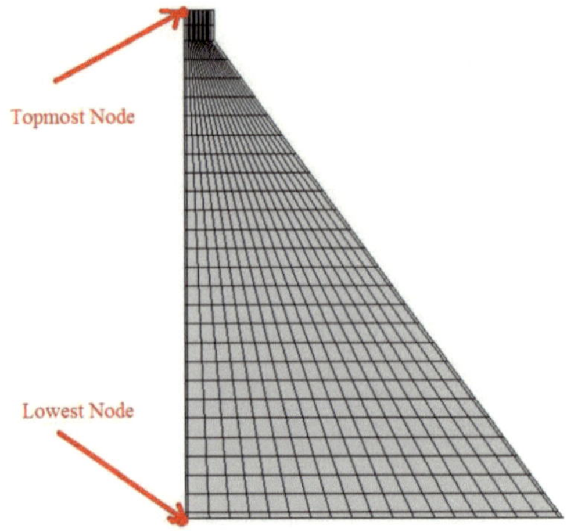

Figure 7.
The location of the topmost node (crest) and the lowest node (heel) at the upstream face.

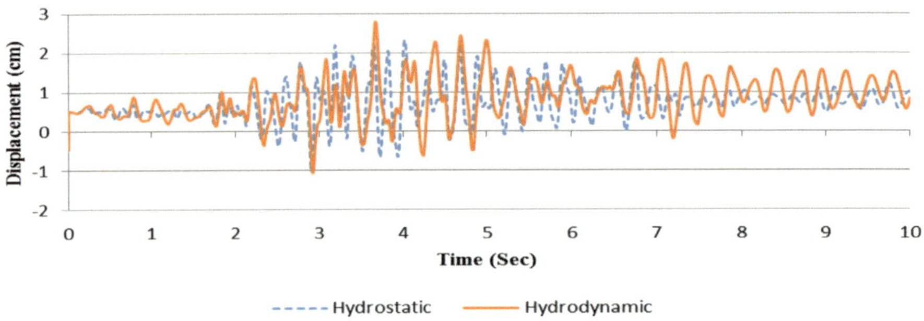

Figure 8.
Relative horizontal displacement of the dam.

displacement can be obtained. For instance in **Figure 10(a)**, 17.30 cm – 14.98 cm = 2.32 cm. This values confirm that the displacement response is increased by 21% when the influence of the hydrodynamic pressure is taken into consideration. In addition to this, the movement is in the positive direction towards the downstream side.

In like manner, the maximum relative vertical displacement of the Kinta dam is shown in **Figure 9**. Based on this figure, the relative vertical displacement of the dam is −0.7 cm at 4.024 second for the hydrostatic effect and − 1.17 cm at 4.391 second for the hydrodynamic water effect. This effect is also shown in **Figures 10(b)** and **11(b)** which are selected at the mentioned seconds. This difference in values again proves the significant effect of hydrodynamic pressure by 67% increase in the vertical direction.

The displacement contours of the Kinta RCC dam in the horizontal and vertical directions due to the Koyna earthquake considering the hydrostatic and hydrodynamic pressure effect are shown in **Figures 10** and **11**. The selected seconds in these figures is chosen accordance with **Figures 8** and **9** in which the maximum relative horizontal and vertical displacement of the RCC dam is occurred. As depicted in **Figure 10(a)**, the nodal horizontal displacement of the dam crest and the heel is 17.30 cm and 14.98 cm, respectively when the hydrostatic pressure is taken into account (17.30–14.98 = 2.32 cm which was described in **Figure 8** as the relative horizontal displacement), whereas, the nodal horizontal displacement of the crest

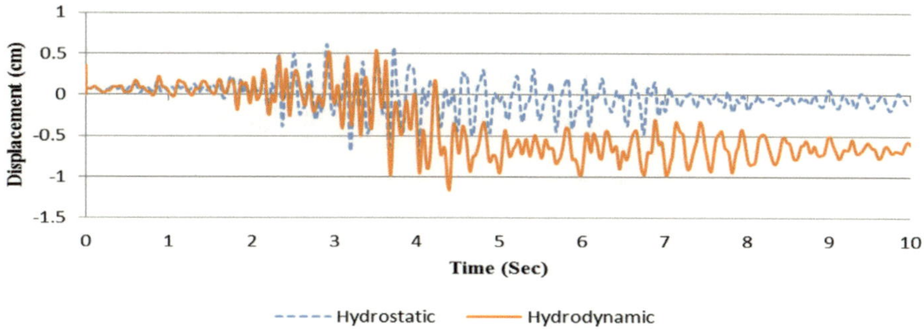

Figure 9.
Relative vertical displacement of the dam.

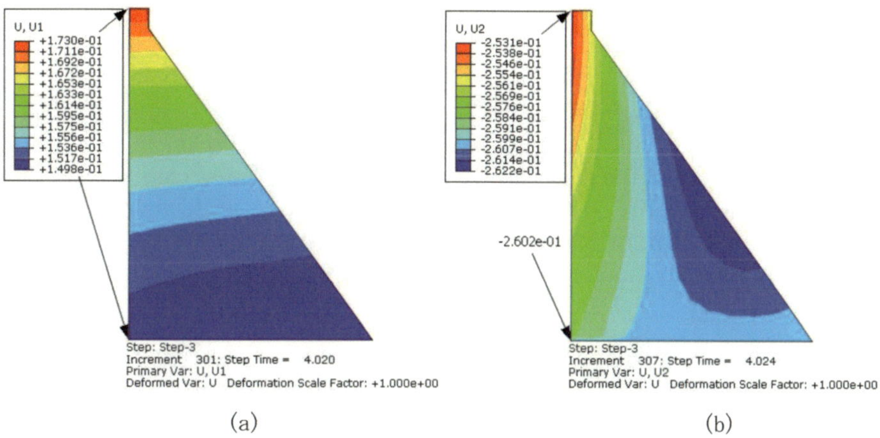

Figure 10.
Displacement (m) contours considering the hydrostatic pressure. (a) Horizontal direction. (b) Vertical direction.

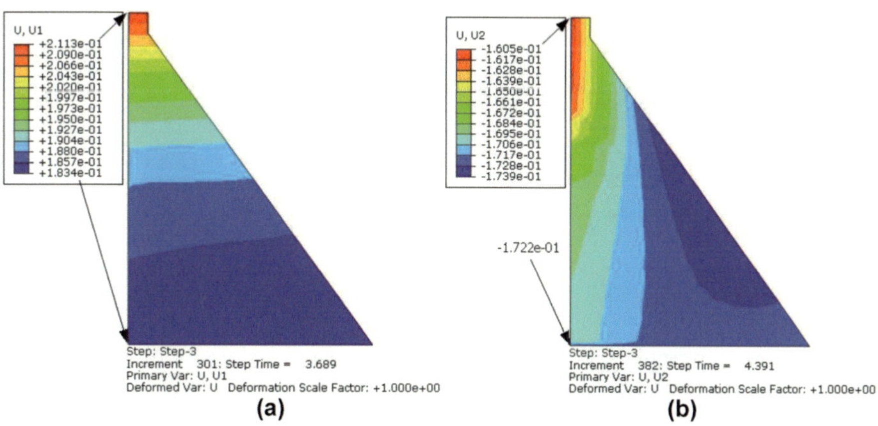

Figure 11.
Displacement (m) contours considering the hydrodynamic pressure. (a) Horizontal. (b) Vertical.

and the heel is 21.13 cm and 18.34 cm considering hydrodynamic water pressure effect (21.13–18.34 = 2.8 cm which was shown in **Figure 8** as the relative horizontal displacement). The comparison of the difference between obtained values shows the hydrodynamic water effect by 22% increase in the horizontal displacement response of the dam. On the other hand, **Figures 10(b)** and **11(b)** illustrate the maximum vertical displacement of the crest and heel nodes of the dam. As depicted in **Figure 10(b)**, the nodal vertical displacement of the dam crest and the heel is 25.31 cm and 26.02 cm, respectively when the hydrostatic pressure is taken into account (25.31–26.02 = −0.7 cm which was described in **Figure 9** as the relative vertical displacement), whereas, the nodal vertical displacement of the crest and the heel is 16.05 cm and 17.22 cm considering hydrodynamic water pressure effect (16.05–17.22 = −1.17 cm which was shown in **Figure 9** as the relative vertical displacement). The comparison of the difference between obtained values shows the hydrodynamic water effect by approximately 36% reduction in the vertical displacement response of the dam. It is because that, the modelled water alongside its assigned weight in the hydrodynamic situation causes smaller vertical movement on both the crest and heel nodes compared to same movements of the aforesaid nodes deliberating the hydrostatic effect on the upstream side.

6.2 Acceleration response

The values of the relative horizontal and vertical acceleration of the dam are indicated in **Figures 12** and **13**. It can be seen from **Figure 12** that, the increase of acceleration response of the dam by 7% from 4 m/s^2 to 4.28 m/s^2 confirms the effect of hydrodynamic pressure on the nonlinear dynamic analysis. However, absorption of acceleration by 2.2 m/s^2 in vertical direction considering the hydrodynamic water effect is approximately similar in compare to the hydrostatic water effect which is only 2.3 m/s^2.

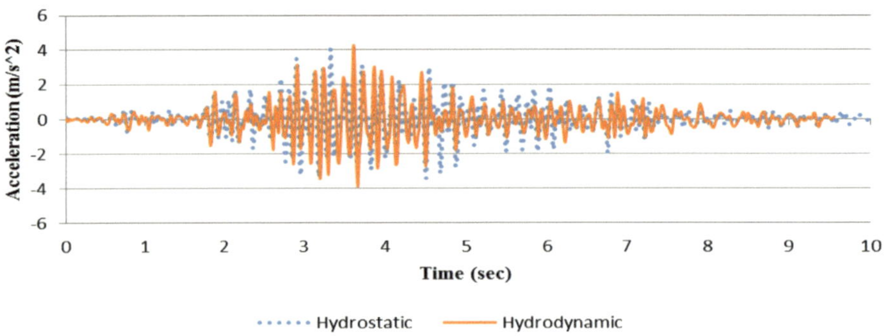

Figure 12.
Relative horizontal acceleration.

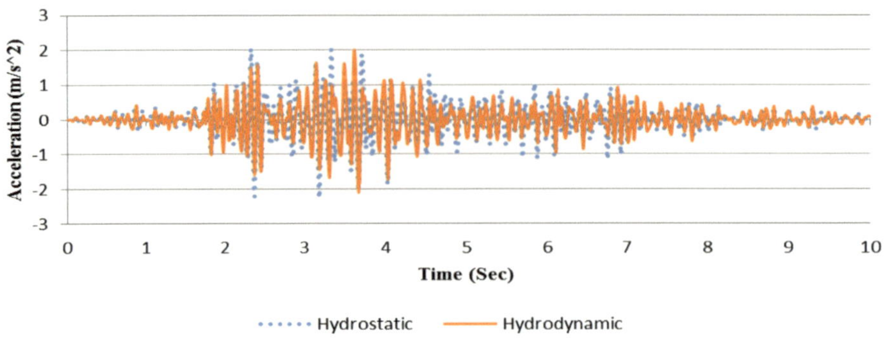

Figure 13.
Relative vertical acceleration.

6.3 Stress

Time history analysis of the maximum principal stresses at the crest and heel elements considering hydrostatic and hydrodynamic pressure effect is plotted in **Figure 14**. It can be seen from the figure that, unlike to the heel element, which experiences the maximum principal stress at initial seconds of the excitations, the crest element experiences no stress at the upstream face in both hydrostatic and hydrodynamic conditions up to second 1.81. Later on, the crest element starts to take an amount of stress and obtains its highest value by 0.195 MPa at 3.19 second under hydrostatic water effect and 0.212 MPa at 3.68 second under hydrodynamic water effect. For the heel element, as shown in **Figure 14(a)** and **(b)**, the stress value is 2.37 MPa at approximately 0.001 second from the earthquake initiation considering both hydrostatic and hydrodynamic pressure effect on the dam. Take note that, the stresses in the heel element after about 5 second is approximately zero. Since during earthquake the stress is changed from an element to another, therefore, after this period the heel element almost does not absorb any serious stress like the other elements. **Figure 15** indicates the maximum principal stresses taken by the heel element (considered as the critical element) at the exact second of absorbing the maximum stress (0.001 second) as expressed above.

Figure 16 indicates the time history analysis of the minimum principal stresses of the crest and heel elements considering hydrostatic and hydrodynamic water pressure. As indicated in the figure, for both the crest and heel elements, the stress

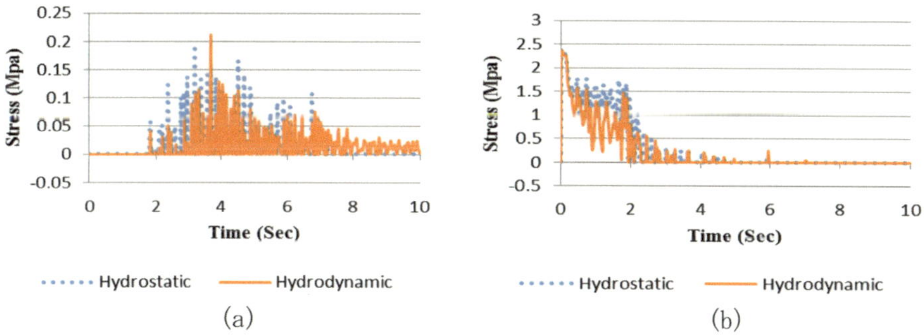

Figure 14.
Time history of the maximum principal stress at the upstream face. (a) Crest element. (b) Heel element.

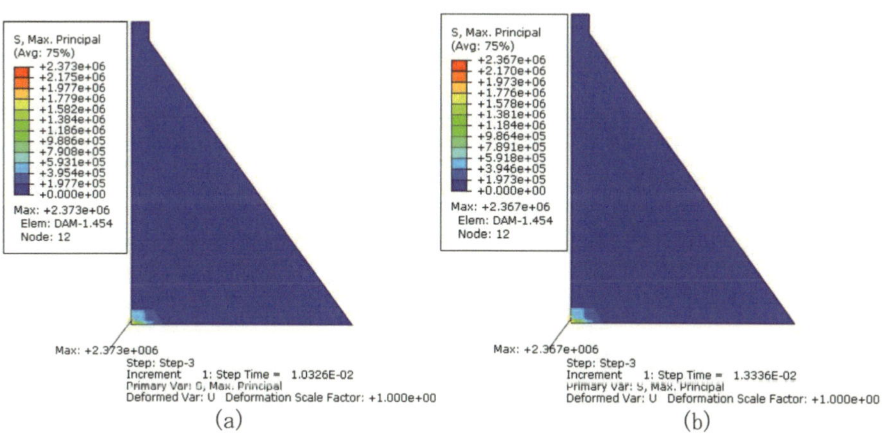

Figure 15.
Maximum principal stress contour of the Kinta RCC dam considering heel element. (a) Hydrostatic effect. (b) Hydrodynamic effect.

is smaller when the hydrodynamic water pressure is deliberated. The minimum principal stress for the crest element is −0.283 MPa considering the hydrostatic effect and − 0.256 MPa considering the hydrodynamic pressure effect as shown in **Figure 16(a)**. Moreover, the time history of the minimum principal stresses attracted by the heel element during the analysis is demonstrated in **Figure 16(b)**. The minimum stress value of the element where the dam interacts with the hydrostatic pressure is −7.84 MPa and − 6.10 MPa in the hydrodynamic interaction condition (22% reduction). **Figure 17(a)** and **(b)** displays the stress contour of the dam at the selected seconds so that the minimum stress values is occurred on the heel element, as explained in **Figure 16(b)**, in both the hydrostatic and hydrodynamic conditions. Apart from the stress values of the heel element in both conditions, the stress propagation inside the dam body is also demonstrated that, the hydrodynamic pressure can influence and change the stress pattern of the dam at different time frame of the analysis.

6.4 Seismic damage of the RCC dam

In the present study, the evaluation of damage level and assessment of the seismic performance of the dam is conducted based on the Concrete Damaged Plasticity (CDP) model. The tensile damage of the considered models is displayed in

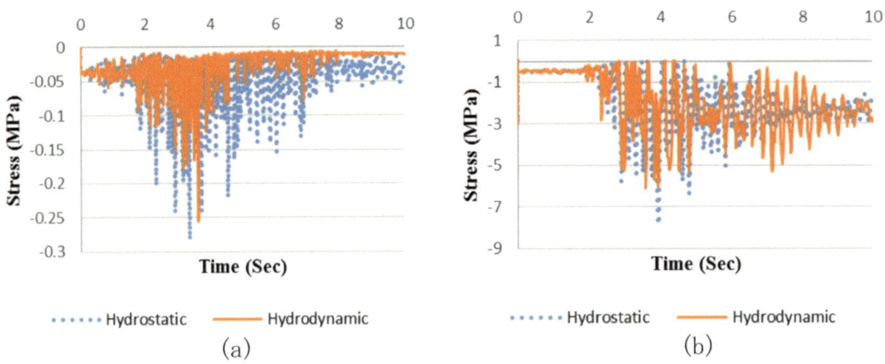

Figure 16.
Time history of the minimum principal stress at the upstream face. (a) Crest element. (b) Heel element.

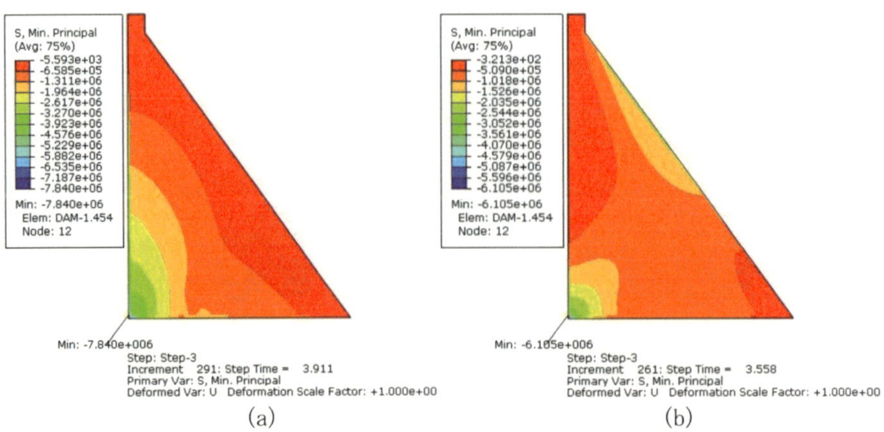

Figure 17.
Minimum principal stress contour of the Kinta RCC dam considering heel element. (a) Hydrostatic effect. (b) Hydrodynamic effect.

Figure 18. Accordance to this figure, the dam in both cases experiences damage (cracking) at the heel elements in the upstream side. Although, the onset of crack pattern is formed in the upstream side at the heel elements, but as the acceleration intensifies, the cracks propagate and develop in different zones of the dam considering the hydrodynamic water effect. **Figure 18(a)** illustrates the crack propagation for the hydrostatic condition which is started from heel zone. When the crack is initiated, it propagates in the horizontal direction toward the downstream side. **Figure 18(b)** gives a picture of tensile damage (cracking) owing to the hydrodynamic effect. As depicted in the figure, the severity of cracking of the dam due to the hydrodynamic pressure effect is not limited to the heel elements but also the dam suffered from additional cracks at the middle zone of the upstream side. This confirms the significance of the nonlinear analysis of the dam considering hydrodynamic pressure effect under seismic ground motions.

Numerous researchers have investigated and reported failure (sliding) mechanism of dams using different approaches [41–45]. It can be concluded from the literature that, there is a direct relationship between opening cracks and sliding (failure) mechanism. Thus, in this study the failure mechanism of the dam and particularly in the heel elements, as the critical elements, considering hydrodynamic pressure effect is presented in **Figure 19**. **Figure 19(a)** shows the crack propagation opf the Kinta RCC dam at the end of the seismic analysis considering hydrodynamic water pressure effect (as is previously illustrated in **Figure 18(b)**). Since, the heel elements at the upstream face are the first elements prone to cracking, therefore, it is tried to investigate the dam failure due to cracking of these elements because of overturning moment owing to the combined effect of the hydrodynamic force and ground motions as shown in **Figure 19(b)**.

According to **Figure 19(a)**, the Kinta RCC dam experiences cracking at the heel elements nearby the upstream face, whilst, the downstream face has no sign of cracking. But, due to the dam overturning and according to description of **Figure 19 (b)**, at the end of the analysis the RCC dam has a movement with magnitude of 12.57 cm (7.65 cm in horizontal direction and 9.97 cm in vertical direction), exactly at the last element of the dam toe. As shown in **Figure 19(a)**, the first six elements are fully cracked and influence the next three elements to be as semi-cracked elements, whereas other 11 elements (elements number 479–489) towards the downstream side are not cracked.

The time history of crack propagation of the heel element (as the critical element) considering hydrostatic and hydrodynamic water pressure effect under the

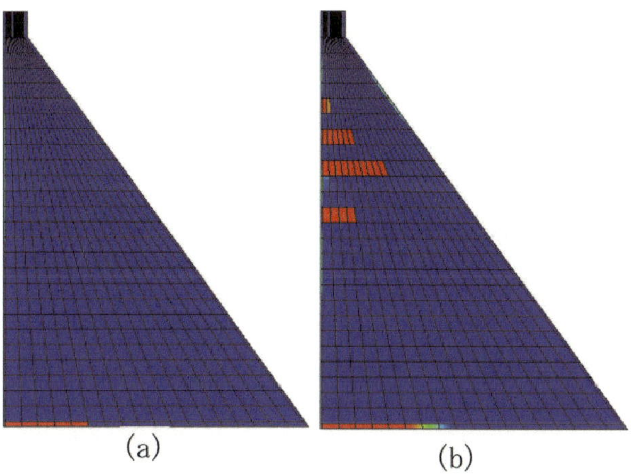

Figure 18.
Tensile damage at the end of analysis. (a) Hydrostatic effect. (b) Hydrodynamic effect.

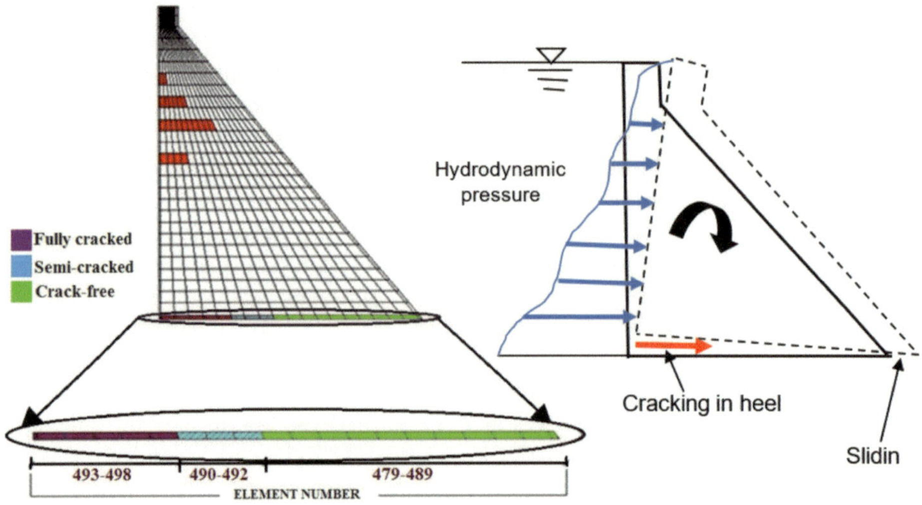

(a) (b)

Figure 19.
Failure (sliding) mechanism of the dam during seismic excitations. (a) selected elements. (b) sliding mechanism.

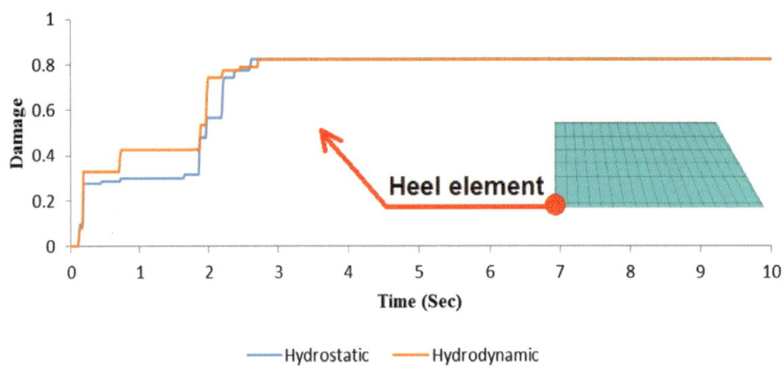

Time

of

Figure 20.
Time history analysis of damage (cracking) development procedure of the selected heel element.

Koyna excitation is indicated in **Figure 20**. The damage range in this figure is provided based on the concrete tension stiffening used in the modeled dam which has direct relation to the tensile strength of the dam concrete as presented in **Table 2**. Based on this explanation and from **Figure 20** it can be concluded that, the heel element until 0.137 second under both hydrostatic and hydrodynamic water pressure effect does not take any damage (cracking). Increasing the reservoir water pressure effect, especially in the hydrodynamic condition, expedites the damage (cracking) procedure of the heel element and leads the dam to experience cracking in the heel zone faster than the condition that the hydrostatic effect is present. Consequently, it is an important matter for the nonlinear seismic analysis of dams to take the hydrodynamic water pressure effect into consideration.

Figure 21 displays the tensile damage process for both the hydrostatic and hydrodynamic conditions during the nonlinear analysis. As given in this figure, some selected times are taken to show the development of cracking inside the dam body under seismic loading. It can be detected from the figure that, from initial time up to 2.73 second the crack propagation trends same pattern. By increasing the

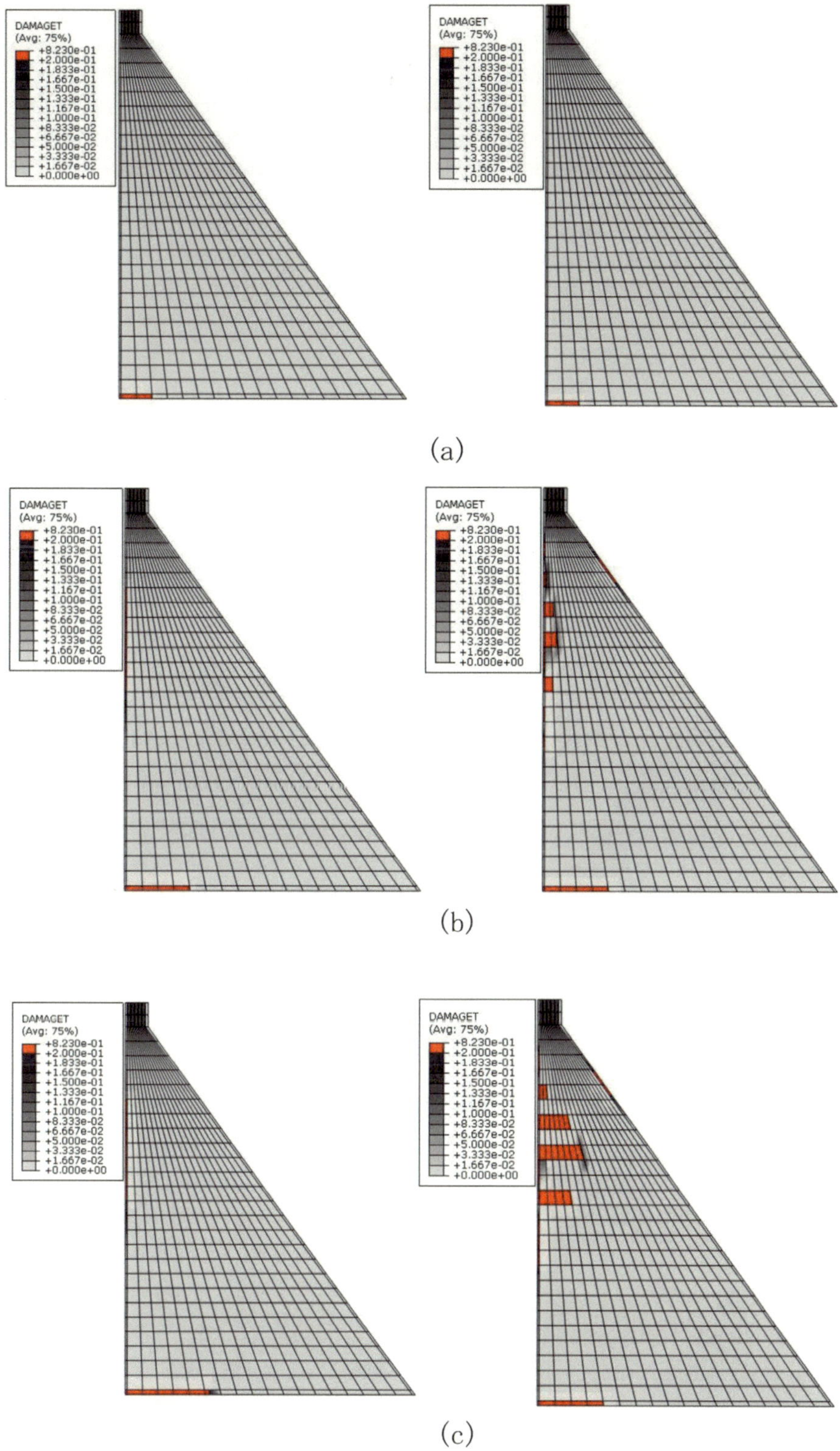

Figure 21.
Tensile damage (cracking) of the Kinta RCC dam at different times considering reservoir hydrostatic (left side) and hydrodynamic (right side) pressure effect. (a) Extension of crack, t = 2.73 sec. (b) Crack propagation at the base affecting more neighboring finite elements, t = 3.67 sec. (c) Severe cracking, t = 4.03 sec.

seismic load and by nearing to the PGA of the Koyna earthquake, at time 3.67 second, the cracks appear with different propagation inside the dam body. At this time it is clear that, the cracking launches at the upstream face towards the downstream direction taking the hydrodynamic pressure effect into consideration. At the time 4.03 second, the orientation of cracking is trended at same level from the heel elements to downstream direction considering the hydrostatic effect, whilst, the hydrodynamic water effect leads the dam to experience severe cracking at the middle zone of the upstream face towards the downstream direction.

Also, **Figure 22** reveals some selected snapshots of the dynamic response of the dam during the analysis to demonstrate the shape deformation of the dam subjected to the seismic motions considering hydrostatic and hydrodynamic pressure effect.

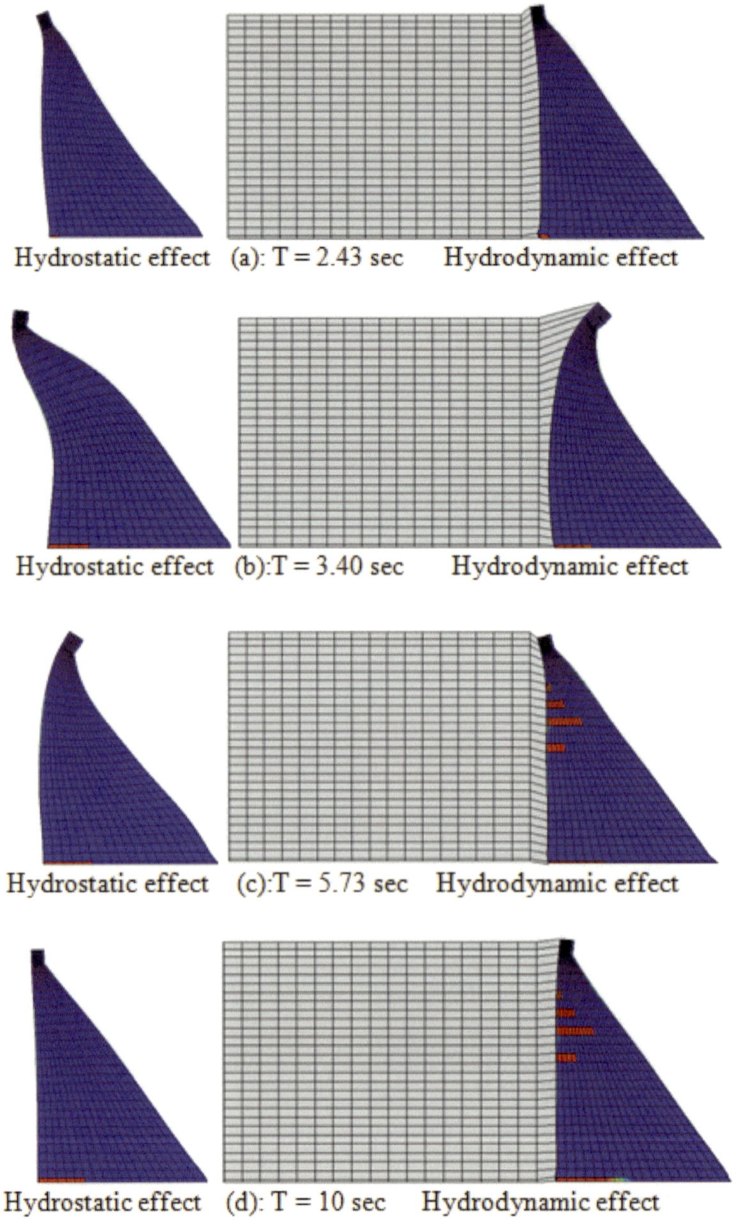

Figure 22.
Snapshots of dam deformation and developed tension cracks in case of both scenarios.

As shown in the figure, different mode shapes at selected times can be observed for the Kinta RCC dam while the hydrostatic and hydrodynamic effect is taken into account. For instance, at the time of 3.40 second and 5.73 second, the mode shapes of the dam is totally dissimilar considering the hydrostatic and hydrodynamic water effect.

7. Conclusions

In this study, an attempt has been made to evaluate the effect of the reservoir hydrostatic and hydrodynamic pressure on the response of RCC dams subjected to earthquake excitations. For this purpose, the Kinta RCC dam has been chosen as a case study and a FEM of the dam-reservoir has been developed. A non-linear seismic analysis of the dam has been conducted in both scenarios. The results of the study have been obtained in terms of acceleration, deformation and displacement, stress and damaged zones. Based on the obtained results, the following conclusions are drawn:

i. The hydrostatic and hydrodynamic pressure exerted at the dam-water interface results higher effects on horizontal than vertical acceleration response.

ii. The nodal displacement of the crest and the heel caused by hydrodynamic effect is increased by 22% in horizontal direction compared to the case when it is omitted. As a result, the relative horizontal displacement of the dam crest is increased by 21% when considering the hydrodynamic pressure effect.

iii. In spite of the fact that, due to considering water weight in hydrodynamic analysis, the nodal displacement of the crest and the heel is decreased by approximately 37% in vertical direction, but the hydrodynamic effect increases the relative vertical displacement of the dam by 67%.

iv. From the maximum and minimum stress analysis, the heel element has the most stress absorption compared to other elements of the dam.

v. Generally, the hydrodynamic effect modifies the deformation shape of the dam during the response and increases the magnitudes of the developed stresses causing more extensive tension crack damages mostly in the heel and upstream zones of the dam.

Eventually, as discussed above, the hydrodynamic pressure highly affects the seismic response and the appearance and extent of the tensile damage zones of the dam. Therefore, for seismic design purposes, this effect has to be seriously taken into consideration.

Author details

Khaled Ghaedi[1,3]*, Farzad Hejazi[2], Meisam Gordan[3], Ahad Javanmardi[1,4], Hamed Khatibi[5] and Ali Joharchi[6]

1 Research and Development Center, PASOFAL Engineering, Kuala Lumpur, Malaysia

2 Faculty of Engineering, Housing Research Center, Universiti Putra Malaysia, Malaysia

3 Department of Civil Engineering, University of Malaya, Kuala Lumpur, Malaysia

4 College of Civil Engineering, Fuzhou University, China

5 Faculty of Engineering, Department of Civil and Environmental Engineering, University of Auckland, New Zealand

6 Department of Civil Engineering, Universiti Kebangsaan Malaysia, Selangor, Malaysia

*Address all correspondence to: khaledqhaedi@yahoo.com

References

[1] K. Ghaedi and Z. Ibrahim, "Earthquake Prediction," in *Earthquakes - Tectonics, Hazard and Risk Mitigation*, T. Zouaghi, Ed. InTechOpen, 2017, pp. 205–227.

[2] K. Ghaedi, Z. Ibrahim, H. Adeli, and A. Javanmardi, "Invited Review: Recent developments in vibration control of building and bridge structures," J. Vibroengineering, vol. 19, no. 5, pp. 3564–3580, Aug. 2017.

[3] A. Javanmardi, Z. Ibrahim, K. Ghaedi, M. Jameel, H. Khatibi, and M. Suhatril, "Seismic response characteristics of a base isolated cable-stayed bridge under moderate and strong ground motions," Arch. Civ. Mech. Eng., vol. 17, no. 2, pp. 419–432, 2017.

[4] M. Gordan, H. A. Razak, Z. Ismail, and K. Ghaedi, "Recent developments in damage identification of structures using data mining," Lat. Am. J. Solids Struct., vol. 14, no. 13, pp. 2373–2401, 2017.

[5] A. Javanmardi, Z. Ibrahim, K. Ghaedi, N. B. Khan, and H. Benisi Ghadim, "Seismic isolation retrofitting solution for an existing steel cable-stayed bridge," *PLoS One*, vol. 13, no. 7, p. e0200482, Jul. 2018.

[6] M. U. Hanif, Z. Ibrahim, K. Ghaedi, H. Hashim, and A. Javanmardi, "Damage assessment of reinforced concrete structures using a model-based nonlinear approach – A comprehensive review," Constr. Build. Mater., vol. 192, pp. 846–865, Dec. 2018.

[7] A. Javanmardi, Z. Ibrahim, K. Ghaedi, H. Benisi Ghadim, and M. U. Hanif, "State-of-the-Art Review of Metallic Dampers: Testing, Development and Implementation," Arch. Comput. Methods Eng., vol. 27, no. 2, pp. 455–478, Apr. 2020.

[8] K. Ghaedi, Z. Ibrahim, A. Javanmardi, and R. Rupakhety, "Experimental Study of a New Bar Damper Device for Vibration Control of Structures Subjected to Earthquake Loads," J. Earthq. Eng., pp. 1–19, Sep. 2018.

[9] M. Gordan, H. Abdul Razaka, Z. Ismail, K. Ghaedi, Z. Tan, and H. Ghayeb, "A hybrid ANN-based imperial competitive algorithm methodology for structural damage identification of slab-on-girder bridge using data mining," Appl. Soft Comput., vol. 88, no. 106013, pp. 56–79, 2020.

[10] Y. Calayir and M. Karaton, "Seismic fracture analysis of concrete gravity dams including dam–reservoir interaction," Comput. Struct., vol. 83, no. 19–20, pp. 1595–1606, Jul. 2005.

[11] W. Guanglun, O. A. Pekau, Z. Chuhan, and W. Shaomin, "Seismic fracture analysis of concrete gravity dams based on nonlinear fracture mechanics," Eng. Fract. Mech., vol. 65, pp. 67–87, 2000.

[12] O. Omidi, S. Valliappan, and V. Lotfi, "Seismic cracking of concrete gravity dams by plastic–damage model using different damping mechanisms," Finite Elem. Anal. Des., vol. 63, pp. 80–97, Jan. 2013.

[13] Z. Shi, M. Nakano, Y. Nakamura, and C. Liu, "Discrete crack analysis of concrete gravity dams based on the known inertia force field of linear response analysis," Eng. Fract. Mech., vol. 115, pp. 122–136, 2014.

[14] G. Wang, Y. Wang, W. Lu, C. Zhou, M. Chen, and P. Yan, "XFEM based seismic potential failure mode analysis of concrete gravity dam–water–foundation systems through incremental dynamic analysis," Eng. Struct., vol. 98, pp. 81–94, 2015.

[15] Q. Xu, J. Y. Chen, J. Li, C. B. Zhang, and C. F. Zhao, "The dynamic

characteristics of the damage probability of a gravity dam," J. Appl. Math. Mech., vol. 79, no. 1, pp. 64–70, 2015.

[16] G. Fenves and A. K. Chopra, "Simplified Earthquake Analysis of Concrete Gravity Dams: Separate Hydrodynamic and Foundation Interaction Effects," J. Eng. Mech., vol. 111, no. 6, pp. 715–735, Jun. 1985.

[17] Malkar. P.F, "Nonlinear response of concrete gravity dams to strong earthquake include ground motion," *Comput. Struct.*, vol. 26(1/2), pp. 165–179, 1987.

[18] M. L. Ayari, "A fracture mechanics based seismic analysis of concrete gravity dams using discrete cracks," *Eng. Fract. Mech.*, vol. 35, no. 1/2/3, pp. 587–598, 1990.

[19] R. Espandar and V. Lotfi, "Comparison of non-orthogonal smeared crack and plasticity models for dynamic analysis of concrete arch dams," Comput. Struct., vol. 81, no. 14, pp. 1461–1474, Jun. 2003.

[20] V. Lotfi and R. Espandar, "Seismic analysis of concrete arch dams by combined discrete crack and non-orthogonal smeared crack technique," Eng. Struct., vol. 26, no. 1, pp. 27–37, Jan. 2004.

[21] M. Akkose, A. Bayraktar, and A. a. Dumanoglu, "Reservoir water level effects on nonlinear dynamic response of arch dams," J. Fluids Struct., vol. 24, no. 3, pp. 418–435, Apr. 2008.

[22] M. Akköse and E. Şimşek, "Non-linear seismic response of concrete gravity dams to near-fault ground motions including dam-water-sediment-foundation interaction," Appl. Math. Model., vol. 34, no. 11, pp. 3685–3700, Nov. 2010.

[23] M. E. Kartal, "Three-dimensional earthquake analysis of roller-compacted concrete dams," Nat. Hazards Earth Syst. Sci., vol. 12, no. 2007, pp. 2369–2388, 2012.

[24] S. Zhang, G. Wang, and W. Sa, "Damage evaluation of concrete gravity dams under mainshock–aftershock seismic sequences," Soil Dyn. Earthq. Eng., vol. 50, pp. 16–27, Jul. 2013.

[25] K. Ghaedi, F. Hejazi, Z. Ibrahim, and P. Khanzaei, "Flexible Foundation Effect on Seismic Analysis of Roller Compacted Concrete (RCC) Dams Using Finite Element Method," KSCE J. Civ. Eng., vol. 22, no. 4, pp. 1275–1287, Apr. 2018.

[26] G. Wang, S. Zhang, C. Zhou, and W. Lu, "Correlation between strong motion durations and damage measures of concrete gravity dams," Soil Dyn. Earthq. Eng., vol. 69, pp. 148–162, 2015.

[27] K. Ghaedi, M. Jameel, Z. Ibrahim, and P. Khanzaei, "Seismic analysis of Roller Compacted Concrete (RCC) dams considering effect of sizes and shapes of galleries," KSCE J. Civ. Eng., vol. 20, no. 1, pp. 261–272, Jan. 2016.

[28] Y. Wang and J. Jia, "Experimental study on the influence of hydraulic fracturing on high concrete gravity dams," Eng. Struct., vol. 132, pp. 508–517, Feb. 2017.

[29] J. Wang, G. Yang, H. Liu, S. Shrawan Nimbalkar, X. Tang, and Y. Xiao, "Seismic response of concrete-rockfill combination dam using large-scale shaking table tests," Soil Dyn. Earthq. Eng., vol. 99, pp. 9–19, Aug. 2017.

[30] "Buletin Ingenieur," *Board of Engineers Malaysia*, vol. 29, no. May 2006, pp. 1–57, 2006.

[31] A. M. Huda, M. S. Jaafar, J. Noorzaei, W. A. Thanoon, and T. A. Mohammed, "Modelling the Effects of Sediment on the Seismic Behaviour of

Kinta Roller Compacted Concrete Dam," Pertanika J. Sci. Technol, vol. 18, no. 1, pp. 43–59, 2010.

[32] GHD, "Study of restrictions on RCC temperature, Stage 2 development of Ipoh water supply," 2002.

[33] US Army Corps of Engineers, *Roller-Compacted Concrete*, no. January 2000. 2000.

[34] A. K. Chopra, *Dynamics of Structures Theory and Applications to Earthquake Engineering*, 4th ed. New Jersey: Prentice-Hall, 2012.

[35] M. U. Hanif, Z. Ibrahim, K. Ghaedi, A. Javanmardi, and S. K. Rehman, "Finite Element Simulation of Damage In RC Beams," J. Civ. Eng. Sci. Technol., vol. 9, no. 1, pp. 50–57, 2018.

[36] S. Zhang and G. Wang, "Effects of near-fault and far-fault ground motions on nonlinear dynamic response and seismic damage of concrete gravity dams," Soil Dyn. Earthq. Eng., vol. 53, pp. 217–229, Oct. 2013.

[37] M. U. Hanif, Z. Ibrahim, M. Jameel, K. Ghaedi, and M. Aslam, "A new approach to estimate damage in concrete beams using non-linearity," Constr. Build. Mater., vol. 124, no. C, pp. 1081–1089, 2016.

[38] K. Ghaedi *et al.*, "Finite Element Analysis of A Strengthened Beam Deliberating Elastically Isotropic And Orthotropic Cfrp Material," *J. Civ. Eng. Sci. Technol.*, vol. 9, no. 2, p. 5, Oct. 2018.

[39] J. Lubliner, J. Oliver, S. Oller, and E. Oñate, "A plastic-damage model for concrete," Int. J. Solids Struct., vol. 25, no. 3, pp. 299–326, 1989.

[40] L. J and F. GL, "Plastic-damage model for cyclic loading of concrete structures," J. Eng. Mech., vol. 124, no. 8, pp. 892–900, 1998.

[41] P. N. Patel, C. C. Spyrakos, and W. Virginia, "Uplifting-sliding response of flexible structures to seismic loads," Eng. Anal. Bound. Elem., vol. 8, no. 4, pp. 185–191, 1991.

[42] M. Ftima and P. Léger, "Seismic stability of cracked concrete dams using rigid block models," Comput. Struct., vol. 84, no. 28, pp. 1802–1814, Nov. 2006.

[43] X. Zhu and O. a. Pekau, "Seismic behavior of concrete gravity dams with penetrated cracks and equivalent impact damping," Eng. Struct., vol. 29, no. 3, pp. 336–345, Mar. 2007.

[44] M. Paggi, G. Ferro, and F. Braga, "A multiscale approach for the seismic analysis of concrete gravity dams," Comput. Struct., vol. 122, pp. 230–238, Jun. 2013.

[45] H. Wang, L. Wang, Y. Song, and J. Wang, "Influence of free water on dynamic behavior of dam concrete under biaxial compression," Constr. Build. Mater., vol. 112, pp. 222–231, 2016.